Advances in Antiferromagnetic Spintronics

Advances in Antiferromagnetic Spintronics

Editor

Atsufumi Hirohata

MDPI • Basel • Beijing • Wuhan • Barcelona • Belgrade • Manchester • Tokyo • Cluj • Tianjin

Editor
Atsufumi Hirohata
Department of Electronic Engineering,
University of York,
Heslington
UK

Editorial Office
MDPI
St. Alban-Anlage 66
4052 Basel, Switzerland

This is a reprint of articles from the Special Issue published online in the open access journal *Magnetochemistry* (ISSN 2312-7481) (available at: https://www.mdpi.com/journal/magnetochemistry/special_issues/advances_antiferromagnetic_spintronics).

For citation purposes, cite each article independently as indicated on the article page online and as indicated below:

LastName, A.A.; LastName, B.B.; LastName, C.C. Article Title. *Journal Name* **Year**, *Volume Number*, Page Range.

ISBN 978-3-0365-3749-8 (Hbk)
ISBN 978-3-0365-3750-4 (PDF)

Contents

About the Editor

Atsufumi Hirohata received his PhD in physics from the University of Cambridge, United Kingdom, in 2001. From 2001 to 2002, he was a Postdoctoral Associate at the Cavendish Laboratory in the University of Cambridge. He then moved to the Francis Bitter Magnet Laboratory in the Massachusetts Institute of Technology, MA, in 2002 as a Postdoctoral Associate. He then served as Researcher at the Department of Materials in Tohoku University, Japan, in 2003 and at the Frontier Research System in RIKEN, Japan, in 2005. He became a Lecturer at the Department of Electronics (now the Department of Electronic Engineering) in the University of York in 2007 and was promoted to Reader, Professor, and then Senior Professor in 2011, 2014, and 2017, respectively. He has edited three books and published more than 160 articles and been responsible for 35 inventions. His research interests include spintronics and magnetic materials. He is the President-Elect of the IEEE Magnetics Society and a member of the Administrative Committee of the Conference on Magnetism and Magnetic Materials. He is a Deputy Editor of Science and Technology of Advanced Materials, an Editor of Journal of Magnetism and Magnetic Materials, SPIN, and Magnetochemistry, an Associate Editor of Frontiers in Physics, and an Editorial Board Member of Journal of Physics D: Applied Physics and SciNotes from 2009 and 2017, and a Guest Professor (Global) at Keio University from 2015 until 2016.

 *magnetochemistry*

MDPI

Editorial

Advances in Antiferromagnetic Spintronics

Atsufumi Hirohata

Department of Electronic Engineering, University of York, York YO10 5DD, UK; atsufumi.hirohata@york.ac.uk

Magnetoresistance (MR) controls signal-to-noise ratios and the corresponding size of conventional spintronic devices [1]. For example, the read head of a hard disk drive (HDD), which has been the most commonly used magnetic storage, decreases the size by improving the MR ratios from a few percent with anisotropic MR (AMR) up to 77% with giant MR (GMR) [2] and to up to 604% with tunnelling MR (TMR) [3] at room temperature. This trend increases its areal recording density due to the reduction in the resulting data bit size. However, the MR ratio has not been improved over the last decade, as shown in Figure 1. This has caused magnetic storages to improve and memories to become slower. In addition, cross-talk between TMR junctions due to the stray fields from their ferromagnetic layers cannot be ignored for further integration. This means alternative materials and/or mechanisms need to be developed for next-generation spintronic devices, especially for storage and memory applications. A strong candidate is antiferromagnetic materials, which do not produce any stray fields.

Citation: Hirohata, A. Advances in Antiferromagnetic Spintronics. *Magnetochemistry* 2022, 8, 37. https://doi.org/10.3390/magnetochemistry8040037

Received: 12 March 2022
Accepted: 15 March 2022
Published: 28 March 2022

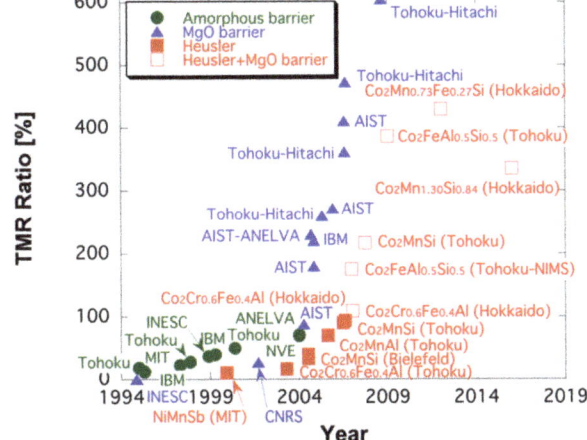

Figure 1. Evolution of TMR ratios at room temperature [4].

Antiferromagnetic materials have been investigated intensively both theoretically and experimentally since their initial discovery by Louis Néel [5]. One of the major applications of antiferromagnets has been to induce interfacial exchange coupling to pin the magnetisation of a neighbouring ferromagnetic layer. This results in a shift in the corresponding magnetisation curve, which can prove the concept of the spin-valve structure [6]. The spin-valve is a basic building block for a HDD read head. Recently, using an electrical current flowing within an antiferromagnetic layer, spin polarisation has been demonstrated to be induced, leading to antiferromagnetic spintronics [7]. For these applications, an $IrMn_3$ alloy has been predominantly used due to its corrosion resistance and robustness against device fabrication processes at the nanometre scale in both the thickness and in-plane dimensions. However, in order to increase the signals of the antiferromagnetic devices, the development of a new material is highly required.

This Special Issue consists of one review and six research articles. The first four articles cover the development of new antiferromagnets for magnetic recording and beyond. Vallejo-Fernandez et al. provided a review on the recent progress in antiferromagnetic films to induce exchange bias onto a neighbouring ferromagnetic film at room temperature [8]. They focused on MnN, achieving the exchange bias of >1 kOe and the anisotropic constant of ~10^6 erg/cm^3. Such a film can offer an alternative to the antiferromagnetic IrMn$_3$ used in a magnetic recording to avoid the use of critical raw materials.

Similar efforts using oxides were made by Shiratsuchi et al. to achieve a large perpendicular exchange bias induced by the magnetoelectric effect in Cr$_2$O$_3$ [9]. The effect can be used to control antiferromagnetic domain states, which can be read out by the magnetisation of the adjacent ferromagnetic layer coupled via the exchange bias induced at their interface. They identified two switching processes: the magnetoelectric field cooling and isothermal modes. The asymmetry yields was reported to be 3.7 ± 0.5 ps/m at 273 K, which is comparable with that of the bulk Cr$_2$O$_3$.

Additionally, Huminiuc et al. grew and characterised polycrystalline Ni$_2$MnAl Heusler alloy films [10]. For the demonstration of room-temperature antiferromagnetism, Fe and Co have been used for partial substitution of Ni. The Fe substitution showed an increase in the magnetic moment with increasing Fe content, while Co substitution can effectively reduce the crystallisation temperature down to 300 °C but with ferromagnetic Co$_2$MnAl segregation. Further compositional optimisation can achieve stoichiometry while maintaining reduced crystallisation in the pseudo-*B*2 phase temperature for antiferromagnetic spintronics.

Ranjbar et al. also reported a large perpendicular exchange energy in rare earth alloys, Tb$_x$Co$_{100-x}$/Cu/[Co/Pt]$_2$ heterostructures [11]. They controlled two competing mechanisms: the effect of Tb content on saturation magnetisation and the coercivity of heterostructures. They demonstrated that the perpendicular exchange energy can be controlled by a Cu interlayer with thicknesses between 0.2 and 0.3 nm up to 1 erg/cm^2 at $x = 24$ and at room temperature. Such a structure can be used in magnetic memory and sensors.

As a new application, magnetisation dynamics in antiferrmagnets were also covered by three articles theoretically and experimentally. Chen et al. demonstrated the manipulation of magnetisation dynamics in the time and frequency domains in a synthetic antiferromagnet using micromagnetic simulations [12]. They found that the time-evolution magnetisations of the two ferromagnets oscillate in-phase at the acoustic mode and out-of-phase at the optic mode. Their simulations confirmed that magnon coupling can be induced in a hybridised resonance mode with a phase difference of up to 90° with respect to the coupling strength. Their method can provide an opportunity to control the magnon interaction in a synthetic antiferromagnet.

Safin et al. discussed a new model for detecting THz frequency signals using antiferromagnetic resonance [13]. The conversion of an electromagnetic signal in THz frequency into a direct current (DC) voltage was calculated and found to be achievable via the inverse spin Hall effect in an antiferromagnet/heavy metal bilayer. Their calculations agreed with an experimentally measured detector sensitivity of 10^{-5}–10^{-6} V/W. The sensitivity can be improved by increasing the magnitude of the bias magnetic field or by decreasing the thickness of the antiferromagnetic layer.

Kim et al. reported the deposition of a crystalline gadolinium iron garnet (GdIG) using a metal organic decomposition method [14]. They demonstrated antiferromagnetic exchange of the rare earth Gd in a ferrimagnetic insulator. For the optimised GdIG films, the magnetic compensation was measured to be at 270 K and the damping constant was measured to be of an order of 10^{-3} based on ferromagnetic resonance measurements. Such a deposition method can offer a high-throughput procedure for ultrafast magnonic applications. Magnons (the quanta of spin waves) can be used to encode information beyond Moore computing applications.

Funding: This work was partially supported by EPSRC (EP/V007211/1 and EP/V047779/1), Royal Society International Exchange Programme and JST CREST (JPMJCR17J5).

Data Availability Statement: Data is contained within the article and available on request with following the guideline set by the University of York (UK).

Conflicts of Interest: The author declares no conflict of interests. The funders had no role in the design of the study; in the collection, analysis, or interpretation of data; in the writing of the manuscript; or in the decision to publish the results.

References

1. Hirohata, A.; Yamada, K.; Nakatani, Y.; Prejbeanu, L.; Diény, B.; Pirro, P.; Hillebrands, B. Review on spintronics: Principles and device applications. *J. Magn. Magn. Mater.* **2020**, *509*, 166711. [CrossRef]
2. Jung, J.W.; Sakuraba, Y.; Sasaki, T.T.; Miura, Y.; Hono, K. Enhancement of magnetoresistance by inserting thin NiAl layers at the interfaces in $Co_2FeGa_{0.5}Ge_{0.5}/Ag/Co_2FeGa_{0.5}Ge_{0.5}$ current-perpendicular-to-plane pseudo spin valves. *Appl. Phys. Lett.* **2016**, *108*, 102408. [CrossRef]
3. Ikeda, S.; Hayakawa, J.; Ashizawa, Y.; Lee, Y.M.; Miura, K.; Hasegawa, H.; Tsunoda, M.; Matsukura, F.; Ohno, H. Tunnel magnetoresistance of 604% at 300 K by suppression of Ta diffusion in CoFeB/MgO/CoFeB pseudo-spin-valves anneal. *Appl. Phys. Lett.* **2008**, *93*, 082508. [CrossRef]
4. Elphick, K.; Frost, W.; Samiepour, M.; Kubota, T.; Takanashi, K.; Sukegawa, H.; Mitani, S.; Hirohata, A. Heusler alloys for spintronic devices: Review on recent development and future perspectives. *Sci. Technol. Adv. Mater.* **2020**, *22*, 235. [CrossRef] [PubMed]
5. Néel, L. Propriétés magnétiques des ferrites. *Ann. Phys.* **1948**, *3*, 137. [CrossRef]
6. Fontana, R.E., Jr.; Gurney, B.A.; Lin, T.; Speriosu, V.S.; Tsang, C.H.; Wilhoit, D.R. Spin Valve Magnetoresistive Sensor with Antiparallel Pinned Layer and Improved Exchange Bias Layer, and Magnetic Recording System Using the Sensor. U.S. Patent 5,701,223, 23 August 1996.
7. Jungwirth, T.; Marti, X.; Wadley, P.; Wunderlich, J. Antiferromagnetic spintronics. *Nat. Nanotech.* **2016**, *11*, 231. [CrossRef]
8. Vallejo-Fernandez, G.; Meinert, M. Recent developments on MnN for spintronic applications. *Magnetochemistry* **2021**, *7*, 116. [CrossRef]
9. Shiratsuchi, Y.; Tao, Y.; Toyoki, K.; Nakatani, R. Magnetoelectric induced switching of perpendicular exchange bias using 30-nm-thick Cr_2O_3 thin film. *Magnetochemistry* **2021**, *7*, 36. [CrossRef]
10. Huminiuc, T.; Whear, O.; Vick, A.J.; Lloyd, D.C.; Vallejo-Fernandez, G.; O'Grady, K.; Hirohata, A. Growth and characterisation of antiferromagnetic Ni_2MnAl Heusler alloy films. *Magnetochemistry* **2021**, *7*, 127. [CrossRef]
11. Ranjbar, S.; Sumi, S.; Tanabe, K.; Awano, H. Large perpendicular exchange energy in $Tb_xCo_{100-x}/Cu(t)/[Co/Pt]_2$ heterostructures. *Magnetochemistry* **2021**, *7*, 141. [CrossRef]
12. Chen, X.; Zheng, C.; Zhou, S.; Liu, Y.; Zhang, Z. Manipulation of time- and frequency-domain dynamics by magnon-magnon coupling in synthetic antiferromagnets. *Magnetochemistry* **2022**, *8*, 7. [CrossRef]
13. Safin, A.; Nikitov, S.; Kirilyuk, A.; Tyberkevych, V.; Slavin, A. Theory of antiferromagnet-based detector of terahertz frequency signals. *Magnetochemistry* **2022**, *8*, 26. [CrossRef]
14. Kim, H.; Van, P.-C.; Jung, H.; Yang, J.; Jo, Y.; Yoo, J.-W.; Park, A.M.; Jeong, J.-R.; Kim, K.-J. Deposition of crystalline GdIG samples using metal organic decomposition method. *Magnetochemistry* **2022**, *8*, 28. [CrossRef]

magnetochemistry

MDPI

Review

Recent Developments on MnN for Spintronic Applications

Gonzalo Vallejo-Fernandez [1],* and Markus Meinert [2],*

1 Department of Physics, University of York, York YO10 5DD, UK
2 Department of Electrical Engineering and Information Technology, Technical University of Darmstadt, 64283 Darmstadt, Germany
* Correspondence: gonzalo.vallejofernandez@york.ac.uk (G.V.-F.); markus.meinert@tu-darmstadt.de (M.M.); Tel.: +44-(0)-1904322265 (G.V.-F.); +49-6151-16-28478 (M.M.)

Abstract: There is significant interest worldwide to identify new antiferromagnetic materials suitable for device applications. Key requirements for such materials are: relatively high magnetocrystalline anisotropy constant, low cost, high corrosion resistance and the ability to induce a large exchange bias, i.e., loop shift, when grown adjacent to a ferromagnetic layer. In this article, a review of recent developments on the novel antiferromagnetic material MnN is presented. This material shows potential as a replacement for the commonly used antiferromagnet of choice, i.e., IrMn. Although the results so far look promising, further work is required for the optimization of this material.

Keywords: antiferromagnetic spintronics; spintronics; exchange bias; MnN; magnetism and magnetic materials

Citation: Vallejo-Fernandez, G.; Meinert, M. Recent Developments on MnN for Spintronic Applications. *Magnetochemistry* **2021**, *7*, 116. https://doi.org/10.3390/magnetochemistry7080116

Academic Editor: Adam J. Hauser

Received: 7 July 2021
Accepted: 30 July 2021
Published: 11 August 2021

Publisher's Note: MDPI stays neutral with regard to jurisdictional claims in published maps and institutional affiliations.

1. Introduction

In 2014, Zhang et al. [1] showed that nontrivial spin Hall effects occur in metallic antiferromagnets such as FeMn, PtMn and IrMn. These results provided evidence that significant spintronic phenomena occur in antiferromagnetic (AF) materials. More recently, Wadley et al. [2] showed that spin–orbit torque could be used to switch CuMnAs electrically. Similarly, Bodnar et al. [3] and Meinert et al. [4] used relativistic Néel spin–orbit torques to switch the Néel vector in Mn_2Au. Spin–orbit torques in AF materials have also been used to switch the magnetization of an adjacent F layer, e.g., [5]. The ability to manipulate the orientation of the AF axes using a spin-polarized current is of huge importance, as AF materials have a relaxation time of $\sim 10^{-12}$ s compared to 10^{-9} s for ferromagnetic (F) materials [6]. Hence, in principle, a switching device based on an AF material would be capable of being many times faster than a conventional Magnetic Random Access Memory (MRAM) device. Such devices would also offer the advantage of not being subject to the demagnetizing field effect present in conventional F devices. AF-based spintronic devices can also be deployed in unconventional computing architectures such as neuromorphic computing, where artificial synapses can be used to execute complex cognitive tasks, e.g., [7]. For these reasons, there has been a significant increase in the level of worldwide activity in the field of AF materials. For a recent review on the topic of AF spintronics, see Reference [8].

The most successful spintronic devices to date are those based on the Giant Magnetoresistance (GMR) effect, i.e., spin-valve, and Tunneling Magnetoresistance (TMR) effect, i.e., tunnel junction, which have formed the basis of the read-head sensor of a Hard Disk Drive (HDD) for decades. AF materials are used in these devices to pin the magnetization of an F (reference) layer via the exchange-bias effect. This effect manifests itself as a shift in the hysteresis loop of the pinned F layer along the field axis. That F layer serves as a magnetic reference in the stack, which is necessary for sensing and MRAM applications. In addition, an enhancement in the coercivity of the material is typically observed. The exchange bias originates from the interplay of interactions across the F/AF interface and the bulk magnetocrystalline anisotropy of the AF. For a review on the topic of exchange bias, see Reference [9]. The first AF material to be used in a GMR commercial device was NiO, which was soon replaced by the metallic alloy FeMn. Although the thermal stability of the devices was significantly improved by doing so, corrosion resistance issues led to the replacement of FeMn by IrMn in the mid-1990s. To this date, IrMn is the material of choice for most industrial applications. PtMn, a simple $L1_0$ alloy, has been used in a thermal MRAM device but not in read heads due to the need for it to be annealed into the $L1_0$ structure [10]. However, Ir is one of the scarcest materials on Earth, making it very expensive. Hence, there is great interest worldwide in the identification of new, usable AF materials that could replace IrMn. The main requirements for such a material are: relatively high anisotropy energy density, which controls its thermal stability; low cost; high corrosion resistance and the ability to induce a significant exchange bias, i.e., loop shift, when grown adjacent to an F layer.

Over the last several years, there have been a number of publications exploring the properties of the equiatomic AF alloy MnN. For instance, broadband ferromagnetic resonance and in-plane angle-dependent measurements have been used to determine the in-plane anisotropies and relaxation of MnN/CoFeB bilayers [11]. One of the attractive properties of this material is that scarcity is not an issue. While there are only $\sim 10^{-5}$ atoms of Ir/million atoms of silicon, Mn is much more abundant ($\sim 10^3$ atoms Mn/million atoms of silicon). Given that almost 80% of the air we breathe is nitrogen, low cost is a given for this compound. In this article, a review of the recent advances in the study of this material will be presented.

2. Manganese Nitride: The Material

Manganese nitride is a complex material that exists in a number of phases, as shown in Figure 1. Depending on the stoichiometry and growth temperature, this compound can exist in a paramagnetic, ferrimagnetic or antiferromagnetic state [12]. Of particular interest to this review article is the tetragonal slightly distorted rocksalt θ phase, θ-MnN, which is AF. From now on, this phase will be referred to simply as MnN. Polycrystalline MnN thin films can be grown by reactive sputtering in a mixed Ar and N_2 atmosphere at temperatures below 673 K and a nitrogen concentration >40% [13]. A 50:50 mixture at a deposition pressure of 2.3×10^{-3} mbar has been proposed as the optimum deposition conditions, although the exact parameters are likely to depend on the deposition system and the details of the sputtering geometry [14].

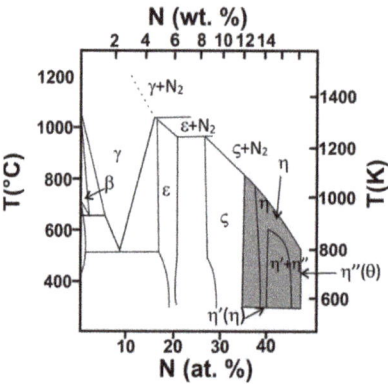

Figure 1. Binary phase diagram of Mn-N. (Redrawn with permission from Journal of Materials Chemistry; published by the Royal Society of Chemistry, 2000) [13].

Neutron powder diffraction experiments have shown that the magnetic order of bulk MnN is collinear AF-I type, i.e., sheet structure with the Mn spins aligned parallel within the c planes [13]. In the same study, the spin orientation was observed to vary as a function of temperature. While at temperatures close to the Néel temperature (~650 K), the moments aligned along the c-axis, at room temperature, the spins were found to be tilted at 23° to the c-axis. These samples were nitrogen deficient. Similar studies on samples where the MnN was saturated suggested that the spin direction was along the c-axis [15]. In both cases, the Mn atoms had a magnetic moment of 3.3 μ_B at room temperature. More recently, a theoretical study has suggested that the spin alignment is along the c-axis [16]. The authors found that the first nearest-neighbor isotropic exchange interactions are AF, while the second nearest-neighbor interactions are strongly ferromagnetic. It was concluded that the interplay between these interactions leads to the AF-I-type ground state. The crystal structure for the θ-MnN phase is shown schematically in Figure 2. It has a tetragonal structure with a lattice constant of a = 4.256 Å and c = 4.189 Å at room temperature resulting in a c/a ratio of 0.984. Interestingly, the easy anisotropy axis lies perpendicular to the long axis of the crystal, as shown in Figure 2. Hence, it might be possible that slight variations in composition might result in modifications to the lattice constant (e.g., increasing the nitrogen content increases the lattice constant) and, potentially, the anisotropy constant of this material. The Néel temperature of this material is slightly higher than 650 K [13], so very similar to that of IrMn, which is ideal for device applications.

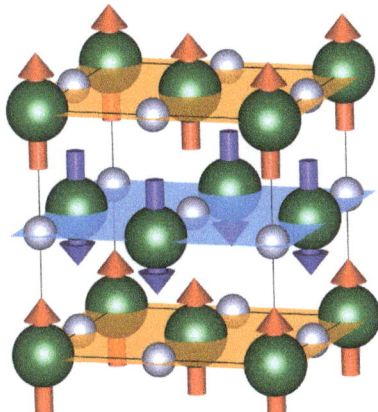

Figure 2. Schematic crystal structure of MnN: green (bigger) spheres correspond to manganese atoms, while smaller, lighter spheres represent nitrogen atoms. The AF-I antiferromagnetic ground state with spin alignment along the c-axis is represented by the arrows (planes) of red and blue (orange and light blue) color.

3. Exchange Bias: In-Plane Studies

In an initial study, Meinert et al. [14] studied the dependence of the exchange bias in MnN/CoFe bilayers at room temperature as a function of the thickness and crystallinity of the AF layer. A typical hysteresis loop from this study is shown in Figure 3 [14].

The crystallinity of the bilayers was controlled both during and after deposition. Samples with composition Ta(10 nm)/MnN(t_{AF})/Co$_{70}$Fe$_{30}$(t_F)/Ta$_2$O$_5$(2 nm) were deposited on thermally oxidized Si wafers. The thickness of the AF layer, t_{AF}, was varied in the range of 6 to 48 nm, while the CoFe thickness, t_F, was varied in the range of 1 to 2.2 nm. After deposition, the samples were annealed at temperatures up to ~600 K in a 6.5 kOe magnetic field. The Ta seed layer grew with (011) orientation and very small grains (~1 nm). Before annealing, the lattice constant of MnN was larger than 4.256 Å, the bulk value reported in the literature. The lattice constant in the film plane was measured as a = 4.10 Å, resulting in a c/a ratio of 1.04. The variation of the loop shift as a function of t_{AF} is shown in Figure 4 (solid red circles). The behavior is identical in shape to that observed for IrMn, whereby no loop shift is observed up to a given value of t_{AF}. This is probably due to a combination of factors. As the thermal stability of the AF grains is increased via the increase in t_{AF}, a sharp onset in the exchange bias is expected. The exchange bias then peaks around a value of t_{AF} of 30 nm, decreasing following a ~1/t_{AF} behavior for thicker films. Following the York model of exchange bias for polycrystalline films, this is a consequence of the shape of the energy barrier to reversal in the AF layer, which is controlled by the distribution of grain volumes, which is lognormal [9]. This is shown schematically in Figure 5. Below a critical volume, V_c, the grains are thermally unstable at the temperature of measurement and do not contribute towards the loop shift. At the higher end of the distribution, there is a second critical volume, V_{set}, above which the grains cannot be set and remain unaligned with the F layer upon annealing. Hence, it is only the grains with volumes between V_c and V_{set} that contribute to the loop shift. Ideally, both critical volumes will be outside the distribution so that the entire AF contributes towards the loop shift. In many cases, as the thickness of the AF layer is increased, a larger fraction of the AF grains cannot be set. Due to the shape of the tail in the lognormal distribution, the exchange bias decreases, mimicking a 1/t_{AF} dependence as mentioned earlier.

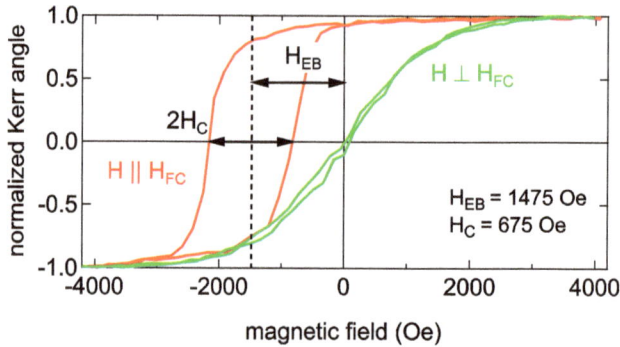

Figure 3. Typical hysteresis loop for a MnN/CoFe exchange bias system. (Reproduced with permission from M. Meinert, Physical Review B; published by the American Physical Society, 2015) [14].

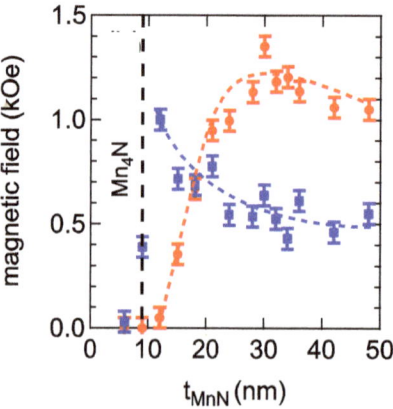

Figure 4. Variation of the exchange bias (solid red circles) and coercive field (blue squares) as a function of MnN thickness. (Reproduced with permission from M. Meinert, Physical Review B; published by the American Physical Society, 2015) [14].

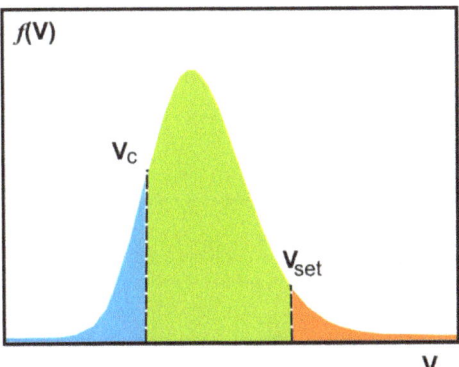

Figure 5. Schematic diagram of the energy barrier to reversal in the AF layer.

A critical difference between the behavior of MnN and IrMn is that the peak in the loop shift as a function of t_{AF} occurs at much lower thicknesses (~8 nm) for the latter. This is probably due to the lower magnetocrystalline anisotropy of MnN (~6 × 10^6 erg/cm^3) compared to that of IrMn (up to ~3 × 10^7 erg/cm^3). A second factor that controls the dependence of the loop shift on t_{AF} is nitrogen desorption upon thermal annealing. In thinner films, nitrogen desorption can lead to a change in the MnN phase to Mn$_4$N, which is not antiferromagnetic, as shown in Figure 4. Thicker films have sufficient nitrogen to withstand desorption effects to higher temperatures. This conclusion was supported by the annealing data, whereby samples with thicker AF layers could withstand higher annealing temperatures before the magnitude of the loop shift was observed to decrease. The dependence of the exchange bias on t_F follows the expected ~1/t_F behavior typical of exchange-bias systems.

The influence of the deposition pressure, annealing temperature and nitrogen content was also investigated [14]. The behavior was quite complex. For instance, while the coercivity was not found to depend greatly on the deposition pressure, the exchange bias was found to decrease almost linearly as the deposition pressure was increased. The annealing temperature dependence was even more complicated. A double 'peak' was observed in the magnitude of the loop shift as a function of the heating temperature. This trend was attributed to an irreversible structural or magnetic transition at the MnN/CoFe interface. In some cases, the annealing temperature was higher than the Néel temperature of the alloy at which point MnN undergoes a recrystallization process which might affect the coupling at the F/AF interface. Again, it is likely that these properties will be (deposition) system dependent.

The thermal stability of the bilayers was assessed via measurement of their median and maximum blocking temperature [14]. Although the thermal stability was lower than that of IrMn-based systems, especially when the thicknesses of the AF layers used in this study are taken into account, a median blocking temperature greater than 373 K was measured for t_{AF} > 15 nm. Importantly, it was found that MnN is rather robust against oxidation, which is critical for device applications. This, coupled with the fact that large exchange-bias values of 1.8 kOe were measured, highlights the potential of MnN for spintronic applications. However, further work is required for the optimization of the growth conditions of this alloy. For instance, controlling the N/Ar mixture during deposition can result in loop shifts in excess of 2.7 kOe at room temperature and an enhanced thermal stability and median blocking temperature > 450 K for compositionally identical structures [17].

The effects of field annealing were further investigated as a function of the thickness of the MnN layer and the field annealing temperature in Ta/MnN/CoFeB exchange-bias systems in a more recent paper [18]. It was found that for thick (48 nm) MnN films, the exchange bias increased due to an improvement in the crystallinity of the films. However, for thinner films (30 nm), the exchange bias was found to decrease and even disappear due to nitrogen migration into the Ta buffer layer, which modified the antiferromagnetic state. Intermixing at the MnN/CoFeB interface was also observed, which was also attributed to the nitrogen deficiency in the MnN layer after annealing in a field. When not enough nitrogen was present, Co and Fe diffused into the MnN, resulting in the reduction of the measured exchange bias. It was suggested that the inclusion of a diffusion barrier layer between the Ta seed layer and the MnN layer might allow for higher annealing temperature to be used without degrading the properties of the films. This is critical if thinner layers were to be used.

Doping has been shown to enhance the thermal stability of MnN/CoFe exchange-bias systems [19]. Undoped samples with structure $Ta(10 nm)/MnN(30 nm)/Co_{70}Fe_{30}/$ (1.6 nm)/Ta(0.5 nm)/$Ta_2 O_5$(5 nm) were deposited on thermally oxidized Si wafers. The samples were annealed at ~600 K in a 6.5 kOe for 15 min. Density Functional Theory (DFT) was used to calculate the defect energies for substitution of one Mn atom in the MnN lattice by a different element. Elements calculated to have both negative and positive defect energies were chosen for this experiment. In particular, Ti, Y, Si, Cr and Fe were selected. Doping elements were added by cosputtering from an RF source. A low power was used to ensure modest doping concentrations. After annealing the crystal structure of the samples was investigated by X-Ray diffraction. The position of the (002) MnN peak was found to shift towards lower angles, indicating a larger lattice constant when elements with a negative defect energy, i.e., Ti, Y, Si and Cr, were used, suggesting that low levels of nitrogen diffusion occurred. On the other hand, when elements with a positive defect energy, i.e., Fe, were used, the (002) peak was shifted towards higher angles when compared to the undoped samples. It was suggested that in this case, significant diffusion occurred because of the weaker binding of nitrogen.

The exchange bias and thermal stability of the doped/undoped samples were investigated as a function of the annealing temperature (273–825 K). The thermal stability of the samples doped with negative defect energies was increased. However, this enhancement was accompanied by a reduction in the magnitude of the loop shift. A similar trend has been observed in IrMn-based exchange-biased systems, whereby the choice of seed layer material has been shown to increase/decrease the anisotropy/loop shift [20]. Modest doping with Ti (3%) or Y (2%) resulted in a loop shift > 1 kOe for annealing temperatures up to ~750 K, as shown in Figure 6, and an enhancement in the thermal stability of 100 K when compared to the undoped sample. Doping with Fe, which has a positive defect energy, showed a negative effect on the thermal stability of MnN/CoFe in agreement with DFT calculations.

Over the last few years, it has become apparent that the Ta buffer layer is key to the performance of MnN-based exchange-bias systems, as it acts as a crystallographic seed layer for the MnN to grow on and a nitrogen sink during the annealing process. The effect of nitrogen diffusion in Ta/MnN/CoFeB stacks has been investigated as a function of the Ta layer thickness [21]. Samples with composition $Ta(t_{Ta})/MnN(30 nm)/Co_{40}Fe_{40}B_{20}(1.6 nm)/Ta$ (0.5 nm)/Ta_2O_5(2 nm) were prepared on thermally oxidized Si wafers. The thickness of the Ta layer, t_{Ta}, was varied in the range of 1 to 15 nm. The samples were annealed at temperatures between 373 and 823 K. While increasing the thickness of the Ta layer improves the crystallinity of the MnN layer, thinner Ta layers provide a smaller nitrogen sink. This leads to a trade-off between thermal stability and large exchange bias. The effect of introducing a TaN_x layer between the Ta seed and the MnN layers was also investigated in the same article [21]. Although the introduction of this extra layer can enhance the thermal stability of the bilayers, it has a negative effect on the magnitude of the loop shift. The data obtained highlighted the complicated nature of MnN, as many interconnected factors can result in structural/magnetic phase transitions. Hence, further work is required in this area to optimize the thermal stability and exchange bias properties of MnN-based exchange bias systems.

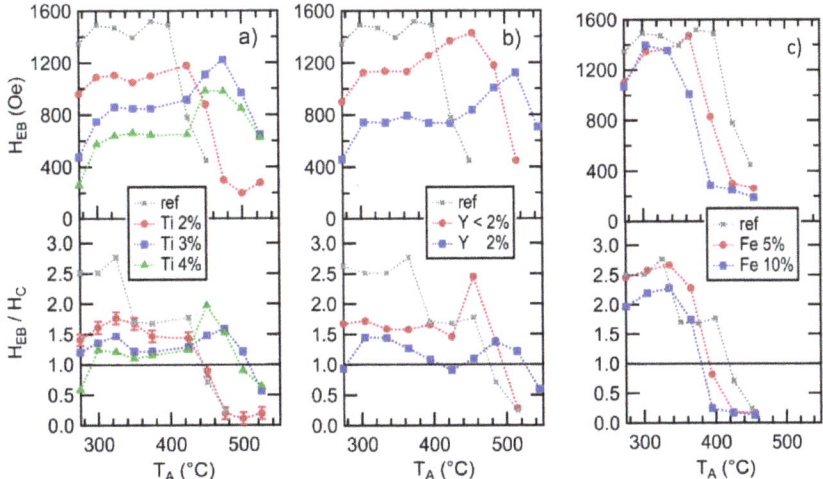

Figure 6. Exchange bias as a function of annealing temperature for different dopant elements: (**a**) Ti, (**b**) Y, (**c**) Fe. The reference measurement (ref) refers to an undoped sample. (Adapted from M. Dunz, Büker and M. Meinert, Journal of Applied Physics 124, 203, 902 (2018); doi:10.1063/1.5051584) [19].

4. Exchange Bias: Out-of-Plane Studies

Magnetic multilayers with perpendicular magnetic anisotropy are also of great interest for spintronic applications given the lower critical current density needed for spin-transfer torque switching when compared to in-plane anisotropy systems, e.g., [22]. Hence, the possibility of achieving a large out-of-plane loop shift using MnN is of great interest. Zilske et al. [23] reported a giant perpendicular exchange bias of 3.6 kOe at room temperature in MnN/CoFeB bilayers. The samples had a similar composition and identical deposition conditions to some of the in-plane samples described earlier. The main difference was in the direction of the applied field used during the annealing process. In this case, the samples were annealed at temperatures in the range of ~400–700 K in a vacuum furnace in the presence of a 6.5 kOe out-of-plane field. This highlights the potential of MnN for integration into perpendicular magnetic tunnel junction/spin valves.

5. Anisotropy Constant

Another key requirement for a given AF material to be considered as a candidate for device applications is that it must have a relatively high magnetocrystalline anisotropy. Sinclair et al. [24] reported the first experimental measurement of the anisotropy constant of MnN in thin-film form from the measurement of the distribution of blocking temperatures in a MnN(t_{AF})/CoFe(2 nm) exchange-bias systems. A detailed description of the technique used can be found in Reference [25]. Briefly, in order to determine K, the samples are initially set at a temperature T_{set} in a positive saturating field for a period of time T_{set}. The samples are then cooled to a temperature, T_{NA}, where the AF layer is free of thermal activation. The field is then reversed so that the F layer is now saturated in the opposite direction. By increasing the temperature of the sample to a temperature T_{act} for a period of time T_{act}, the AF grains are progressively reversed. The samples are then cooled to T_{NA}, and the hysteresis loop is measured. By increasing the value of T_{act}, the energy barrier distribution within the AF is mapped. As a result, the hysteresis loop shifts from negative to positive field values. The critical point for the measurement of K is the value of T_{act} at which the exchange bias goes to 0. This is commonly known as the median blocking temperature as, at the point, half of the AF grains are oriented in the original setting direction and half of the grains in the opposite direction. Assuming the grain volume distribution within the sample is known, the value of K can be calculated using the median grain volume following that antiferromagnets are subject to thermal activation following a Néel–Arrhenius law. Figure 7 shows the blocking temperature distribution for samples of varying MnN thicknesses. From these data, it was concluded that MnN has an anisotropy energy density of 6.3×10^6 erg/cc as compared to a maximum anisotropy constant of 3.2×10^7 erg/cc for IrMn [20]. However, thicknesses > 20 nm appear to be necessary to achieve thermal stability above room temperature due to the small grain sizes (~5 nm). If the lateral grain size of the films could be increased by tuning the deposition conditions, it seems reasonable to suggest that the thickness of the MnN could be reduced, making it a very suitable candidate for device applications.

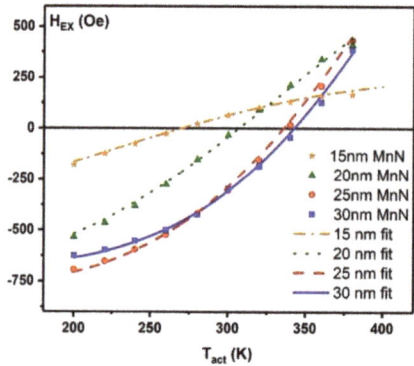

Figure 7. Reverse field cooling experiment as a function of the MnN thickness. (Reproduced with permission from G. Vallejo-Fernandez, Journal of Magnetism and Magnetic Materials; published by Elsevier, 2019) [24].

6. Electrical Switching

The low anisotropy energy of MnN grains, especially at low thicknesses, can in some cases be seen as an advantage. The spin–orbit torque-induced electrical switching of polycrystalline MnN layers with the spin Hall effect of Pt has recently been studied [26]. The electrical manipulation of the magnetic order with current pulses was observed over a broad temperature range between 160 K and 260 K. With increasing temperature, a more efficient switching of the magnetic order was observed, with a simultaneous reduction in the relaxation time for randomization of the magnetic order. The analysis of the magnetocrystalline anisotropy in Reference [24] and the quantitative analysis of the temperature-dependent relaxation dynamics both consistently point to an energy barrier of 0.5 eV to 0.7 eV for the grains that participate in the observed switching. This renders the magnetic order of MnN easily switchable with the spin Hall effect but at the expense of a rapid thermal relaxation back to a disordered equilibrium. The same grains that are only weakly blocked or unblocked in the switching experiment do not contribute to exchange bias because of their relaxation dynamics.

7. Conclusions

Over the last few years, MnN has been demonstrated to show potential as a new AF material for spintronic applications. A summary of the properties of this compound compared to other Mn-based AFs is given in Table 1. This material can exist in a tetragonal structure, which by its very nature creates an anisotropy. One of the main advantages of this alloy compared to conventionally used AF materials is its low cost. Significant loop shifts, similar to values obtained for IrMn-based systems, have been achieved both for in-plane and out-of-plane MnN-based exchange-bias systems. However, the growth conditions and postdeposition treatment(s) for this compound are yet to be optimized. Large AF thicknesses are still required to observe the desired magnetic properties. It has also become apparent that the Ta buffer layer commonly used is critical to control the crystallinity of the MnN layer and the level of nitrogen diffusion observed upon thermal annealing. It might be possible that other materials offer better performance than Ta, and/or high-temperature deposition of the buffer layer might result in an increase in the crystallinity of the adjacent MnN layer. Tuning the deposition conditions of the MnN layer might result in an increase in the lateral grain size of the films, which would allow for a reduction in the thickness of the MnN layer. Furthermore, it is feasible that slight variations in composition might result in modifications to the lattice constant of the MnN films and, potentially, the anisotropy constant of the material. The mutually consistent results from the exchange bias and electrical switching studies corroborate our understanding of the importance of thermal activation in both phenomena in granular antiferromagnetic systems. While MnN is a novel but already well-studied antiferromagnet, open questions remain. How does the exchange bias behave in highly crystalline, epitaxial MnN thin films? What is the easy axis or easy plane of the polycrystalline MnN films, and how does it influence the observed exchange bias in detail? Thus, it seems like there is much more work to be done on MnN before our understanding of this intriguingly complex material is complete.

Table 1. Summary of the main properties of Mn-based AF materials.

Material	Crystal Structure	Cost	Typical Exchange Bias (kOe)	Anisotropy Constant (erg/cm^3)
FeMn	fcc	Low	<1	~10^5 [27]
IrMn	fcc	High	>1	~10^7 [20]
PtMn	L1$_0$	High	<1	~10^7 [28]
MnN	fct	Very low	>1	~10^6 [24]

Author Contributions: Both authors have contributed to the preparation of this manuscript. All authors have read and agreed to the published version of the manuscript.

Funding: This research was partly funded by the Royal Society No. INF\R1\180070.

Institutional Review Board Statement: Not applicable.

Informed Consent Statement: Not applicable.

Data Availability Statement: Data can be made available upon request.

Acknowledgments: G.V.F. would like to thank K. O'Grady for careful reading of the manuscript and insightful comments.

Conflicts of Interest: The authors declare no conflict of interest.

References

1. Zhang, W.; Jungfleisch, M.B.; Jiang, W.; Pearson, J.E.; Hoffmann, A.; Freimuth, F.; Mokrousov, Y. Spin Hall effects in metallic antiferromagnets. *Phys. Rev. Lett.* **2014**, *113*, 196602. [CrossRef]
2. Wadley, P.; Howells, B.; Elezny, J.; Andrews, C.; Hills, V.; Campion, R.P.; Novak, V.; Olejnik, K.; Maccherozzi, F.; Dhesi, S.S.; et al. Electrical switching of an antiferromagnet. *Science* **2016**, *351*, 587–590. [CrossRef]
3. Bodnar, S.Y.; Mejkal, L.; Turek, I.; Gomonay, O.; Sinova, J.; Sapozhnik, A.A.; Elmers, H.-J.; Klaui, M.; Jourdan, M. Writing and reading antiferromagnetic Mn 2 Au by Néel spin-orbit torques and large anisotropic magnetoresistance. *Nat. Commun.* **2018**, *9*, 1–7. [CrossRef]
4. Meinert, M.; Graulich, D.; Matalla-Wagner, T. Electrical switching of antiferromagnetic Mn$_2$Au and the role of thermal activation. *Phys. Rev. Appl.* **2018**, *9*, 064040. [CrossRef]
5. Borders, W.A.; Akima, H.; Fukami, S.; Moriya, S.; Kurihara, S.; Horio, Y.; Sato, S.; Ohno, H. Analogue spin–orbit torque device for artificial-neural-network-based associative memory operation. *Appl. Phys. Express* **2017**, *10*, 013007. [CrossRef]
6. Vallejo-Fernandez, G.; Aley, N.P.; Chapman, J.N.; O'Grady, K. Measurement of the attempt frequency in antiferromagnets. *Appl. Phys. Lett.* **2010**, *97*, 22505. [CrossRef]
7. Fukami, S.; Ohno, H. Perspective: Spintronic synapse for artificial neural network. *J. Appl. Phys.* **2018**, *124*, 151904. [CrossRef]

8. Baltz, V.; Manchon, A.; Tsoi, M.; Moriyama, T.; Ono, T.; Tserkovnyak, Y. Antiferromagnetic spintronics. *Rev. Mod. Phys.* **2018**, *90*, 015005. [CrossRef]
9. O'Grady, K.; Fernandez-Outon, L.E.; Vallejo-Fernandez, G. A new paradigm for exchange bias in polycrystalline thin films. *J. Magn. Magn. Mater.* **2010**, *322*, 883–899. [CrossRef]
10. Farrow, R.F.C.; Marks, R.F.; Gider, S.; Marley, A.C.; Parkin, S.S.P.; Mauri, D. Mn_xPt_{1-x}: A new exchange bias material for Permalloy. *J. Appl. Phys.* **1997**, *81*, 4986–4988. [CrossRef]
11. Rai, A.; Dunz, M.; Sapkota, A.; Zilske, P.; Mohammadi, J.B.; Meinert, M.; Mewes, C.; Mewes, T. Unidirectional and uniaxial anisotropies in the MnN/CoFeB exchange bias system. *J. Magn. Magn. Mater.* **2019**, *485*, 374–380. [CrossRef]
12. Gokcen, N.A. The Mn-N (Manganese-Nitrogen) system. *Bull. Alloy. Phase Diagr.* **1990**, *11*, 33. [CrossRef]
13. Leineweber, A.; Niewa, R.; Jacobs, H.; Kochelmann, W. The manganese nitrides η-Mn3N2 and θ-Mn6N5+ x: Nuclear and magnetic structures. *J. Mater. Chem.* **2000**, *10*, 2827–2834. [CrossRef]
14. Meinert, M.; Büker, B.; Graulich, D.; Dunz, M. Large exchange bias in polycrystalline MnN/CoFe bilayers at room temperature. *Phys. Rev. B* **2015**, *92*, 144408. [CrossRef]
15. Suzuki, K.; Yamaguchi, Y.; Kaneko, T.; Yoshida, H.; Obi, Y.; Fujimori, H.; Morita, H.J. Neutron diffraction studies of the compounds MnN and FeN. *Phys. Soc. Jpn.* **2001**, *70*, 1084–1089. [CrossRef]
16. Simon, E.; Yanes, R.; Khmelevskyi, S.; Palotás, K.; Szunyogh, L.; Nowak, U. Magnetism and exchange-bias effect at the MnN/Fe interface. *Phys. Rev. B* **2018**, *98*, 094415. [CrossRef]
17. Dunz, M.; Schmalhorst, J.; Meinert, M. Enhanced exchange bias in MnN/CoFe bilayers after high-temperature annealing. *AIP Adv.* **2018**, *8*, 056304. [CrossRef]
18. Quaterman, P.; Hallsteinsen, J.; Dunz, M.; Meinert, M.; Arenholz, A.; Borchers, J.A.; Grutter, J. Effects of field annealing on MnN/CoFeB exchange bias systems. *Phys. Rev. Mater.* **2019**, *3*, 064413. [CrossRef]
19. Dunz, M.; Büker, B.; Meinert, M. Improved thermal stability in doped MnN/CoFe exchange bias systems. *J. Appl. Phys.* **2018**, *124*, 203902. [CrossRef]
20. Aley, N.P.; Vallejo-Fernandez, G.; Kroeger, R.; Lafferty, B.; Agnew, J.; Lu, Y.; O'Grady, K. Texture effects in IrMn/CoFe exchange bias systems. *IEEE Trans. Magn.* **2008**, *44*, 2820–2823. [CrossRef]
21. Dunz, M.; Meinert, M. Role of the Ta buffer layer in Ta/MnN/CoFeB stacks for maximizing exchange bias. *J. Appl. Phys.* **2020**, *128*, 153902. [CrossRef]
22. Nishimura, N.; Hirai, T.; Koganei, A.; Ikeda, T.; Okano, K.; Sekiguchi, Y.; Osada, Y. Magnetic tunnel junction device with perpendicular magnetization films for high-density magnetic random access memory. *J. Appl. Phys.* **2002**, *91*, 5246–5249. [CrossRef]
23. Zilske, P.; Garulich, D.; Munz, D.; Meinert, M. Giant perpendicular exchange bias with antiferromagnetic MnN. *Appl. Phys. Lett.* **2017**, *110*, 192402. [CrossRef]
24. Sinclair, J.; Hirohata, A.; Vallejo-Fernandez, G.; Meinert, M.; O'Grady, K. Thermal stability of exchange bias systems based on MnN. *J. Magn. Magn. Mater.* **2019**, *476*, 278–283. [CrossRef]
25. Vallejo-Fernandez, G.; Fernandez-Outon, L.E.; O'Grady, K. Measurement of the anisotropy constant of antiferromagnets in metallic polycrystalline exchange biased systems. *Appl. Phys. Lett.* **2007**, *91*, 212503. [CrossRef]
26. Dunz, M.; Matalla-Wagner, T.; Meinert, M. Spin-orbit torque induced electrical switching of antiferromagnetic MnN. *Phys. Rev. Res.* **2020**, *2*, 013347. [CrossRef]
27. Fernandez-Outon, L.E.; Vallejo-Fernandez, G.; Manzoor, S.; Hillebrands, B.; O'Grady, K. Interfacial spin order in exchange biased systems. *J. Appl. Phys.* **2008**, *104*, 093907. [CrossRef]
28. Kato, T.; Ito, H.; Sugihara, K.; Tsunashima, S.; Iwata, S. Magnetic anisotropy of MBE grown MnPt3 and CrPt3 ordered alloy films. *J. Magn. Magn. Mater.* **2004**, *272*, 778–779. [CrossRef]

MDPI

Article

Magnetoelectric Induced Switching of Perpendicular Exchange Bias Using 30-nm-Thick Cr_2O_3 Thin Film

Yu Shiratsuchi *, Yiran Tao, Kentaro Toyoki and Ryoichi Nakatani

Department of Materials Science and Engineering, Graduate School of Engineering, Osaka University, Osaka 565-0871, Japan; yiran.tao@mat.eng.osaka-u.ac.jp (Y.T.); toyoki@mat.eng.osaka-u.ac.jp (K.T.); nakatani@mat.eng.osaka-u.ac.jp (R.N.)
* Correspondence: shiratsuchi@mat.eng.osaka-u.ac.jp

Abstract: Magnetoelectric (ME) effect is a result of the interplay between magnetism and electric field and now, it is regarded as a principle that can be applied to the technique of controlling the antiferromagnetic (AFM) domain state. The ME-controlled AFM domain state can be read out by the magnetization of the adjacent ferromagnetic layer coupled with the ME AFM layer via exchange bias. In this technique, the reduction in the ME layer thickness is an ongoing challenge. In this paper, we demonstrate the ME-induced switching of exchange bias polarity using the 30-nm thick ME Cr_2O_3 thin film. Two typical switching processes, the ME field cooling (MEFC) and isothermal modes, are both explored. The required ME field for the switching in the MEFC mode suggests that the ME susceptibility (α_{33}) is not deteriorated at 30 nm thickness regime. The isothermal change of the exchange bias shows the hysteresis with respect to the electric field, and there is an asymmetry of the switching field depending on the switching direction. The quantitative analysis of this asymmetry yields α_{33} at 273 K of 3.7 ± 0.5 ps/m, which is comparable to the reported value for the bulk Cr_2O_3.

Keywords: magnetoelectric effect; antiferromagnetism; Cr_2O_3 thin film; exchange bias

Citation: Shiratsuchi, Y.; Tao, Y.; Toyoki, K.; Nakatani, R. Magnetoelectric Induced Switching of Perpendicular Exchange Bias Using 30-nm-Thick Cr_2O_3 Thin Film. *Magnetochemistry* **2021**, *7*, 36. https://doi.org/10.3390/magnetochemistry7030036

Academic Editor: Atsufumi Hirohata

Received: 18 February 2021
Accepted: 5 March 2021
Published: 9 March 2021

Publisher's Note: MDPI stays neutral with regard to jurisdictional claims in published maps and institutional affiliations.

1. Introduction

Controlling antiferromagnetic (AFM) domain has been an active pursuit because of the possible applications, such as ultrahigh density storage and THz devices. The difficulty is mainly in the control and detection of the AFM domain state because no net magnetization emerges from AFM materials. So far, some techniques to control the AFM domain state have been proposed, such as spin orbit torque [1]. The magnetoelectric (ME) effect, an induction of magnetization (M) by an electric field (E) or an induction of electric polarization (P) by a magnetic field (H), is also one root. The ME effect appears in some insulating antiferromagnets as a result of the simultaneous breakings of time- and spatial-inversion symmetries. The strength of the ME effect is quantified by the ME susceptibility α (= dM/dE = dP/dH). The linear ME effect was experimentally observed in the bulk Cr_2O_3 crystal in the early of 1960s [2,3]. In 1966, Martin and Anderson revealed, based on the symmetrical argument, that the sign of α depends on the orientation of Néel vector and the AFM domain state was consequently controllable [4]. Now, the linear ME effect of Cr_2O_3 has been recognized renewably as the ferroic feature in the presence of the finite E or H [5], and it was confirmed by the Cr_2O_3 thin film [6].

For the detection of the AFM domain state, the usage of exchange bias at the ferromagnetic (FM)/AFM interface [7] or the anomalous Hall effect (AHE) in the heavy metal such as Pt on AFM [8] has been proposed. The device architecture based on the former scenario was proposed by Chen et al., as ME-random access memory (ME-RAM) [9]. The former scenario is based on the fact that the ME-controlled AFM domain state is detectable via the FM magnetization with the assumption that the exchange bias polarity is coupled with the AFM domain state. The detection of ME-controlled Cr_2O_3 domain state was first demonstrated using the bulk Cr_2O_3 substrate [7] and it has been developed to all-thin-film system with the Cr_2O_3 layer [10,11]. Notably, the above prerequisite has also been proven experimentally by means of the element-specific magnetic domain observation [12]. In this approach, the large output signal is expected so that the output voltage is determined by the FM magnetization direction. Another approach based on the AHE of the heavy metal has the benefit that the stacking structure is very simple and the switching energy can be reduced compared with the former scenario because the interfacial exchange coupling with the FM spin is absent. Instead, the output voltage could be small. At the first stage for both attempts, the Cr_2O_3 thickness is high, typically above 200 nm [10,11,13–15], and the reduction in the Cr_2O_3 thickness is an ongoing demand. Until now, the switching of the exchange bias polarity was realized using the 50-nm thick Cr_2O_3 [16] and the AHE detection was confirmed down to 20-nm thick Cr_2O_3 regime [17]. In particular, the former scheme has a prerequisite that the exchange bias has to maintain the low Cr_2O_3 thickness. It was reported that the critical thickness of the appearance of the exchange bias was relevant to the AFM domain wall width [18]. For the case of Cr_2O_3, the AFM domain wall width is reported to be 20–60 nm depending on the lattice deformation [19]. Hence, it is important to investigate the applicable thickness of former scheme below 50 nm. In this paper, we explored the reduction in Cr_2O_3 thickness in the former approach and demonstrate the ME-induced switching of the exchange bias polarity using the 30-nm thick Cr_2O_3 layer.

There are two typical ME field application processes: ME field cooling (MEFC) and isothermal switching. In the previous reports, mainly the ME field cooling (MEFC) process was done. The isothermal mode was not as much because of the difficulty despite its importance for practical use. This is partly because for the isothermal switching, the high dielectric resistance is required because of the high required ME energy compared with the MEFC mode [15,20]. In this paper, we present both types of switching. Based on the required field condition, we show that the ME susceptibility (α) is not deteriorated in the 30-nm-thickness regime.

2. Materials and Methods

Pt 2 nm/Co 0.25 nm/Au 1.0 nm/Cr_2O_3 30 nm/Pt 20 nm stacked film was prepared on an α-Al_2O_3(0001) substrate. The film preparation was done by the DC magnetron sputtering system with the base pressure below 1×10^{-6} Pa. The 20-nm-thick Pt layer was deposited at 873 K on the ultrasonically cleaned substrate as a buffer layer to align the crystallographic orientation of the Cr_2O_3 layer. The Pt-buffer layer also works as the bottom electrode to apply E to the Cr_2O_3 layer. The Cr_2O_3 was formed by sputtering of a pure Cr target in Ar + O_2 gas mixture at the substrate temperature of 773 K. The 1.0-nm thick Au layer was used to tune the strength of the interfacial exchange coupling J_{INT} between Co and Cr_2O_3 [21]. Unless the suitable spacer layer was inserted, the exchange bias cannot be maintained in the temperature regime where the ME susceptibility is high [15]. The Co and Pt top layers were deposited at room temperature. The 2-nm thick Pt layer prevents the oxidization of ultrathin FM Co layer and also acts as the induction of the perpendicular magnetic anisotropy. The crystallographic orientation of each layer was characterized by using a reflection high-energy electron diffraction (RHEED). The RHEED chamber is directly connected to the sputtering chamber, and hence the RHEED observations could be done without exposing the sample to air. As shown in Figure 1, the RHEED pattern on the Cr_2O_3 layer is streaky, which shows the flat surface. The diffraction pattern indicates that the Cr_2O_3 layer grows with the c-axis along the growth direction and that the twin boundary is included along the [11$\bar{2}$0] direction.

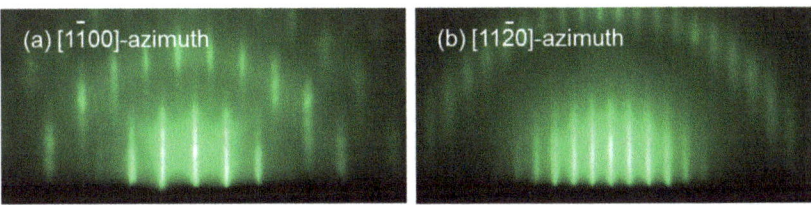

Figure 1. Reflection high-energy electron diffraction RHEED images of Cr_2O_3 layer with (a) [$1\bar{1}00$] and (b) [$11\bar{2}0$] azimuth.

X-ray reflection (XRR) measurement was carried out to confirm the well-defined stacking structure. The XRR profile shown in Figure 2a shows the clear oscillation above $2\theta/\omega$ of 14°, which proves the sharp interfaces and the well-defined stacking structure. The FFT analysis of the oscillation gives the actual Cr_2O_3 thickness as 31.6 nm. In the high-angle X-ray diffraction (XRD) profile shown in Figure 2b, any diffraction peaks other than Pt(111) and α-Al_2O_3(0001) substrate were not observed. Many Laue fringes were observed around the Pt(111) diffraction peaks (Figure 2c), which indicate the good crystalline quality of the Pt buffer layer. It should be note that the diffraction peaks from the Cr_2O_3 layer are overlapped with those from the Pt(111)/Pt(222) diffraction peaks. The diffractions from the Cr_2O_3 layer could be masked because of the larger atomic scattering factor of Pt compared with Cr^{3+} and O^{2-}.

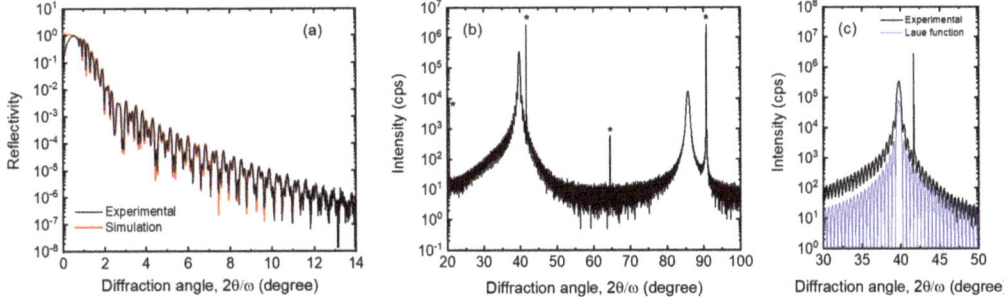

Figure 2. (a) X-ray reflection (XRR), (b) high-angle $2\theta/\omega$ profile of the film. (c) represents the enlarged profile of (b) around $2\theta/\omega$ = 30-50°. Black and red lines in (a) represent the experimental result and the fitted result, respectively. Blue line in (c) represent the calculated profile using the Laue function.

Magnetic characterizations were carried out mainly by means of the AHE measurements after applying the ME field. For the AHE measurements, the film was patterned into the microdot with 15 μm diameter. On top of the microdot, the four top electrode (Cr/Au) was prepared by the lift-off technique. The optical microscope image with the electrical circuit is shown in Figure 3a. In our sample structure, the current flows in the Pt/Co/Au layer wherein the AHE detects the Co magnetization perpendicular to the film plane. This is suitable in this work because the studied film has the perpendicular magnetic anisotropy. In this work, the AHE loops were measured as a function of H applied to the direction perpendicular to the film. From the AHE loop, the exchange bias field, H_{ex}, a shift of the magnetization curve along the H axis was evaluated. H_{ex} was estimated as $\Delta H_C/2$ where H_C is the switching field (coercivity) for up-to-down and down-to-up switching of the FM magnetization. E was applied between top (Pt/Co/Au) and bottom (Pt-buffer layer) electrodes, i.e., the direction perpendicular to film. The positive directions of H and E are defined as the direction from bottom (substrate side) to top (film side) of the film.

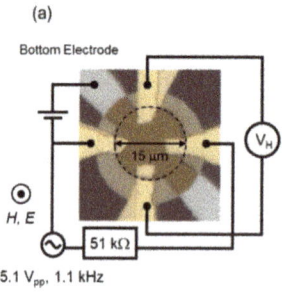
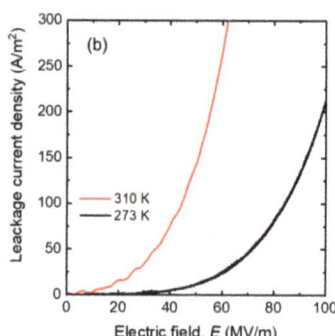

Figure 3. (**a**) Optical microscope image of the device used for the test and the electric circuitry. (**b**) *I-V* curves of the Cr_2O_3 layer measured at 310 K (red) and 273 K (black).

Here, we denote the ME field as the simultaneously applied *H* and *E* because the energy gain by the ME effect ΔF is expressed as

$$\Delta F = \alpha_{ij} \cdot E_i H_j \tag{1}$$

Together with the crystallographic characterization of Cr_2O_3, the ME susceptibility evaluated in this paper is α_{33}. As mentioned above, there are two typical processes to apply to the ME field: MEFC and isothermal modes. In the MEFC mode, the sample was once heated to above the Néel temperature of Cr_2O_3 (~307 K for bulk Cr_2O_3 [22]), 310 K. Then, H_{MEFC} ($\mu_0 H_{MEFC}$ = 5, 6, 7 T) and *E* (0–47 MV/m) were applied. In maintaining both fields, the sample was cooled to the AHE measurement temperature, 250 K at which the exchange bias polarity was checked. After measuring the AHE loop, the above heating and cooling processes were repeated with the different ME field condition. More details of the MEFC process can be found in [7,12].

In the isothermal process, the temperature kept constant during the ME field application and the AHE measurements. In this process, the sample was cooled from 310 K to 273 K under the *H* ($\mu_0 H$ = 0.6 T) application during the cooling: the conventional field-cooling (FC) to induce the exchange bias. At the constant temperature (273 K), $H_{isothermal}$ ($\mu_0 H_{isothermal}$ = 6 T) and *E* (−60 – +100 MV/m) were simultaneously applied typically for 30 s. After that, *E* was removed and the AHE as a function of *H* measured. In the isothermal process, the sequential ME field application is important so that the exchange bias switching occurs accompanied by the hysteresis. In this work, the ME fields were applied in the following sequences.

- ME field application: for example, $\mu_0 H_{isothermal}$ = 6 T and *E* = +60 MV/m;
- Removing *E* and AHE measurement as a function of $\mu_0 H$;
- ME field application: for example, $\mu_0 H_{isothermal}$ = 6 T and *E* = +80 MV/m;
- Removing *E* and AHE measurement as a function of $\mu_0 H$;

These processes were repeated with increasing *E* until the exchange bias polarity is fully switched. Finally, after the exchange bias polarity was switched, the sign of the ME field was reversed, and the similar processes were repeated with the negative *E*.

- ME field application: for example, $\mu_0 H_{isothermal}$ = 6 T and *E* = 0 MV/m;
- Removing *E* and AHE measurement as a function of $\mu_0 H$;
- ME field application, for example, $\mu_0 H_{isothermal}$ = 6 T and *E* = −18 MV/m;
- Removing *E* and AHE measurement as a function of $\mu_0 H$;
- ME field application: for example, $\mu_0 H_{isothermal}$ = 6 T and *E* = −23 MV/m;
- Removing *E* and AHE measurement as a function of $\mu_0 H$;

These processes were repeated with decreasing *E* until the exchange bias polarity is fully switched again. Details of the isothermal switching protocol can be found in our previous report [20]. Note that for both processes, the highest $\mu_0 H$ during the AHE loops was ± 700 mT, which is low enough to switch the exchange bias polarity by H alone ($|\mu_0 H| > 8$ T) [23].

3. Results and Discussions

Prior to showing the results on the ME-induced switching, it is helpful to show the electric resistance of the Cr_2O_3 layer. Using the same device used for the AHE measurements, the electric resistance was measured. Typical I-V curves measured at 310 K and 273 K are shown in Figure 3b. The I-V curves show the non-linear increase with respect to the voltage as is generally observed in an insulator thin film. Even at the most severe conditions adopted in this work (at highest temperature and at the highest E adopted), the current density is in the range of 10^2 A/m^2, which is low enough compared with the current-induced magnetization switching, such as the spin-orbit torque mechanism, typically above 10^9 A/m^2 [24,25].

First, we show the switching of the exchange bias polarity by the MEFC process. Figure 4a shows the series of AHE loops measured after the MEFC with $\mu_0 H_{MEFC} = 7$ T. When the electric field was not applied, e.g., the conventional FC, the negative exchange bias of $\mu_0 H_{ex} = -108$ mT appears (black curve). The similar AHE loops were obtained for E below 16 MV/m (blue curve). With increasing E, the step at about +100 mT starts to be observed in the AHE loop, a signature of the appearance of positive exchange bias. The two-step AHE loops are observed in the E range of 24–43 MV/m. The two-step magnetization curve is attributed to the co-existence of the positive and negative exchange biased domains. The similar two-step exchange biased state was observed in the previously reported ME-induced switching [16] and the magnetization curve after zero-field cooling [26]. The positive exchange bias grows with increasing E, the step at about -100 mT suppresses and instead, that at about +100 mT enhances. Finally, above $E = 47$ MV/m, the AHE loop shows the full positive exchange bias. In Figure 4b, the change in $\mu_0 H_{ex}$ as a function of E during the MEFC is shown. We find that the change in the exchange bias is common to every $\mu_0 H_{MEFC}$. It should be noted that the AHE loops show the tiny hysteresis, i.e., the low coercivity. This is attributed to the usage of the Au spacer layer instead of the Pt spacer layer as in [21]. The Au spacer layer is suitable to tune the interfacial magnetic anisotropy in maintaining high exchange bias with suppressing the coercivity enhancement. The details of the role on the spacer layer can be found in our previous paper [21].

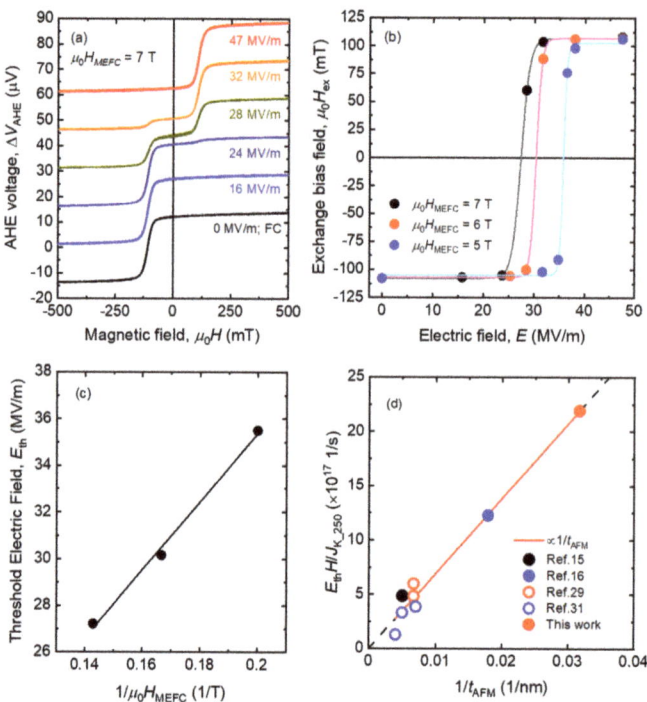

Figure 4. (a) Anomalous Hall effect (AHE) loops as a function of $\mu_0 H$ after the magnetoelectric field cooling (MEFC). The E value for the MEFC are 0 (black), 16 MV/m (blue), 24 MV/m (dark blue), 28 MV/m (yellow), 32 MV/m (orange), and 47 MV/m (red). For every case, $\mu_0 H_{MEFC}$ was 7 T. **(b)** Change in the exchange bias field with E. Black, red and blue point correspond to the case of $\mu_0 H_{MEFC} = 7$ T, 6 T and 5 T. Lines represent the fitted results using Equations (2) and (3). **(c)** Change in the threshold E with $1/\mu_0 H_{MEFC}$. Line represents the linearly fitted result. **(d)** $1/t_{AFM}$ dependence of the required EH product to switch the exchange bias. The EH values are normalized by the exchange anisotropy energy density at 250 K (see text).

The growth of the positive exchange-biased state by E is understood by the energy competition between J_{INT} and the energy gain by the ME effect (Equation (1)) [27]. The former is caused by the interfacial exchange coupling between FM (Co) and interfacial AFM (Cr) spins. Previously, we reported that Co and interfacial Cr spins couples antiferromagnetically; the spin orientation of Co and interfacial Cr is opposite [28]. Under the positive H_{MEFC} adopted in this paper, upward Co spin and downward Cr spin are the favorable spin alignments near the interface. When E during the MEFC is weak, this effect predominantly determines the interfacial spin alignment and yields the negative exchange-biased state. Conversely, the energy gain by the ME effect expressed by Equation (1) favors the upward Cr spin orientation favoring the positive exchange-biased state, which competes J_{INT}. The phenomenological expression of this energy competition is given by [27]

$$\Delta G = \left[(\alpha_{33} E + M_{AFM}) H_{MEFC} - \frac{J_{INT}}{t_{AFM}} \right] V_{AFM} \qquad (2)$$

where J_{INT} denotes the interfacial exchange coupling energy (J/m²), t_{AFM} is the AFM layer thickness (m), M_{AFM} is the uncompensated AFM moment in the Cr_2O_3 layer (Wb/m²) and V_{AFM} is the activation volume (m³) In the MEFC process, the spin configuration and the consequent exchange-biased state are determined during the cooling. Hence, the thermal fluctuation is also taken into account. Assuming that the fraction of the negative and positive exchange-biased state obeys the Boltzmann distribution, the change in the exchange bias should be expressed as [27]

$$\mu_0 H_{ex}(E) \propto \tanh\left(\frac{\Delta G}{k_B T}\right) \tag{3}$$

where k_B is a Boltzmann constant (J/K) and T is an absolute temperature (K). In Figure 4b, the fitted results using Equation (3) are shown as the solid lines. The results for all adopted MEFC conditions follow the same mathematical form. When H_{MEFC} decreases, the curve shifts toward the high E. This shift can be also understood based on Equation (2). For the weaker H_{MEFC}, the higher E is required to compensate J_{INT}/t_{AFM}. According to this argument, the required E to switch the exchange bias should be inversely proportional to H_{MEFC}. Defining the threshold E, E_{th} to switch the exchange bias as E at which $\mu_0 H_{ex}$ becomes zero, we plot the E_{th} as a function of $1/\mu_0 H_{MEFC}$ in Figure 4c. The E_{th} linearly increases with $1/H$, which is in agreement with the above argument that the driving force for switching derives from the energy difference shown in Equation (1). The slope of Figure 4c gives the required EH product, which has been used as a measure of the ME-induced switching [15,27,29], as 1.24×10^{14} V·A/m^2.

In other multiferroic systems such as BiFeO$_3$, the cross-correlation coefficient such as the piezoelectric coefficient (d_{33}) can be deteriorated with decreasing thickness because of the lattice confinement by the epitaxial strain [30]. Here, we discuss the deterioration/enhancement of α_{33} with reducing t_{AFM} by comparing the obtained value with the previous reports for the similar FM/Cr$_2$O$_3$ stacked system [15,16,29,31]. To compare the EH product, J_{INT} in Equation (2) has to be taken into account. Although the direct evaluation of J_{INT} is difficult, the exchange anisotropy energy density J_K (= $H_{ex} \cdot M_S \cdot t_{FM}$) can be used as a measure of J_{INT}; in the simple pinned spin model [32] or the weak J_{INT} limit in the domain wall model [33], J_{INT} and J_K become equal to each other. Because J_K depends on the temperature, J_K measured at 250 K, J_{K_250} is used as a measure of J_{INT} as in the previous paper [16]. To evaluate J_K, the saturation magnetization per unit area, $M_S \cdot t_{FM}$ was measured based on the magnetization curve (M-H curve). The $M_{S\_FM} \cdot t_{FM}$ value is $(5.7 \pm 0.7) \times 10^{-10}$ Wb/m, which is higher than the bulk Co because of the sizable spin polarization of Pt and Au attached with Co [21]. In Figure 4d, the relationship between $E_{th} \cdot H_{MEFC}/J_K$ and t_{AFM} is shown. As expected from Equation (2), $E_{th} \cdot H_{MEFC}/J_K$ roughly proportional to $1/t_{AFM}$. Notably, the $E_{th} \cdot H/J_K$–$1/t_{AFM}$ relationship is maintained up to the 30-nm-thickness regime, which suggests that α_{33} does not deteriorate in this thickness regime. This finding is in agreement that the switching is not highly influenced by the strain effect discussed above.

We further evaluate the α_{33} value based on the isothermal switching mode. Figure 5a shows the AHE loops after applying the ME fields, $\mu_0 H_{isothermal}$ = 6 T and E = +98 MV/m (red curve) and $\mu_0 H_{isothermal}$ = 6 T and E = −41 MV/m (blue curve), which correspond to the positive and the negative exchange-biased states, respectively. We find that the switching between two states reversibly occurs. Figure 5b shows the change in $\mu_0 H_{ex}$ as a function of E. The clear hysteresis is observed, which indicates the presence of the energy barrier to switch the exchange bias polarity. We also find that the hysteresis shifts along the E-axis toward the positive direction, resembling the exchange bias in the M-H curve.

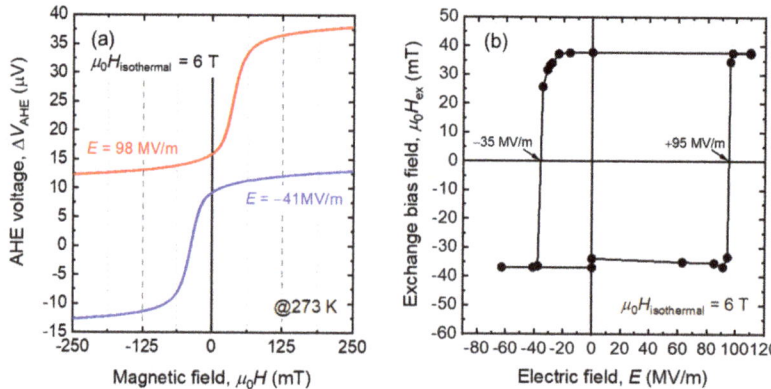

Figure 5. (a) AHE loops as a function of $\mu_0 H$ after the applying the magnetoelectric (ME) fields in an isothermal way. Blue and red curves represent the AHE loop after applying $E = +98$ MV/m and -41 MV/m with $\mu_0 H_{\text{isothermal}} = 6$ T, respectively. (b) Change in the exchange bias field as a function of E with $\mu_0 H_{\text{isothermal}} = 6$ T. In (b), the threshold electric field is shown by text.

We analyze the shift of the hysteresis based on the phenomenological expression of the switching energy assuming the coherent rotation of the AFM spin. The switching ME condition is expressed as [20]

$$(\alpha_{33} E_{th} + M_{AFM}) H = -2K_{AFM} \mp \frac{J_{INT}}{t_{AFM}} \quad (4)$$

where K_{AFM} denote the magnetic anisotropy energy density of the AFM layer (J/m^3). In this expression, the change in the sing of α_{33} was taken into account. The sing of the second term depends on the switching direction: the negative-to-positive switching and the positive-to-negative switching. According to Equation (4), the asymmetry of the coercive E in the $\mu_0 H_{\text{ex}}$-E hysteresis is relevant to the difference in the switching energy caused by the unidirectional nature of J_{INT}. Hence, the shift of the $\mu_0 H_{\text{ex}}$-E hysteresis is essentially same as J_K. J_K is quantified as

$$J_K = \frac{M_{S\_FM} \cdot t_{FM} \cdot \Delta H_C}{2} = \frac{\alpha_{33} H \cdot t_{AFM} \cdot \Delta E_{th}}{2} \quad (5)$$

The first expression is the difference in the Zeeman energy for up-to-down and down-to-up switching of the FM magnetization. Values of $M_{S\_FM} \cdot t_{FM}$, $(5.7 \pm 0.7) \times 10^{-10}$ Wb/m, and the exchange bias field $H_{\text{ex}} = \Delta H_C/2$ obtained from the AHE loop yields J_K at 273 K of 0.017 ± 0.002 mJ/m^2. Because the exchange bias polarity is determined by the interfacial AFM spin direction [12], the $\mu_0 H_{\text{ex}}$-E hysteresis represents that of the interfacial AFM spin as a function of E. The second expression relies on this fact. Using J_K, $\mu_0 H_{\text{isothermal}}$ (6 T), t_{AFM} (31.6 nm, determined by XRR, see above) and ΔE_{th} (95−35 = 60 MV/m), α_{33} is yielded as 3.7 ± 0.5 ps/m. This value is in good agreement with the reported values at the same temperature for the bulk Cr_2O_3 and 500-nm-thick Cr_2O_3 film [6], ~4 ps/m.

One may imagine that the magnetic anisotropy energy of Cr_2O_3 can be evaluated from the K_{AFM} value. In our experiments, the exchange bias was checked at zero E, i.e., after removing the ME field. Besides, the exchange bias polarity is determined by the interfacial AFM spin direction [12]. Considering them, the change in the exchange bias reflects the interfacial antiferromagnetic spin direction at the remanent state after applying the ME field. Hence, the physical meaning of the $\mu_0 H_{\text{ex}}$-E curve is similar to the remanent magnetization curve for the interfacial antiferromagnetic spin. The remanent magnetization curve is often used in the field of magnetic recording/storage and the details of the remanent magnetization curve can be found in [34]. Equation (4) assumes the coherent rotation whereas the switching occurs with the AFM domain wall motion [12]. In this case, the K_{AFM} evaluated using Figure 5b and Equation (4) corresponds the ME energy equivalent to the remanent coercive E, which is typically two or three orders lower than the magnetocrystalline anisotropy energy density. This is similar to the fact that the switching H (the coercivity) in the *M-H* loop is different from the magnetic anisotropy field, $2K_{FM}/M_{S\_FM}$ of the FM layer (K_{FM} is a magnetic anisotropy energy density of FM layer).

Finally, we discuss the role of M_{AFM} and the approach to decrease the switching energy using M_{AFM}. As shown in Figure 4d, the reduction in t_{AFM} enhances the switching energy, and this tendency is common to the isothermal mode. According to Equations (2) and (4), if M_{AFM} is parallel to $\alpha_{33}H$, it would assist the switching. M_{AFM} can be attributed to the finite magnetization at the bulk site [35,36] and/or the interfacial uncompensated moment [20]. The former relies on the defect-induced magnetization [35] and/or the selective substitution of the non-magnetic element such as Al to one sub-lattice [36]. Although the reduction effect using M_{AFM} has been actually reported for the MEFC mode [28], the validity for the isothermal mode has not been proven, which will be investigated in the near future.

4. Conclusions

In summary, we investigate the ME-induced switching of the exchange bias, using especially the Cr_2O_3 thin film with the 30-nm thickness regime. While the previous reports seem to pay less attention to the isothermal reversible switching because of the difficulty, this switching mode as well as the MEFC-driven switching are demonstrated. Based on the argument on the switching energy, the non-deterioration of the ME susceptibility α_{33} in the studied thickness regime is found. The quantitative analysis based on the asymmetric switching energy in the isothermal switching gives the similar α_{33} value to the reported one for the bulk Cr_2O_3. While the cross-correlation functionality can deteriorate with decreasing thickness in some multiferroic materials, the finding in this paper could be a light for the further reduction in the ME-based all-thin-film system, which may be useful in spintronic applications.

Author Contributions: Conceptualization, Y.S. and R.N.; sample fabrication, Y.T.; data curation, Y.T. and K.T.; formal analysis, Y.S., Y.T. and K.T.; writing—original draft preparation, Y.S.; writing—review and editing, R.N.; funding acquisition, Y.S. All authors have read and agreed to the published version of the manuscript.

Funding: This research was partly funded by JSPS KAKENHI, grant number 19H00825.

Data Availability Statement: The study did not report any data.

Conflicts of Interest: The authors declare no conflict of interests. The funders had no role in the design of the study; in the collection, analysis, or interpretation of data; in the writing of the manuscript, or in the decision to publish the results.

References

1. Wadley, P.; Howell, B.; Železený, J.; Andrews, C.; Hills, V.; Campion, V.; Novák, V.; Olejník, K.; Maccherozzi, F.; Dhesi, S.; et al. Electrical switching of an antiferromagnet. *Science* **2016**, *351*, 587–590. [CrossRef] [PubMed]
2. Astrov, D.N. The magnetoelectric effect in antiferromagnetics. *Sov. Phys. JETP* **1960**, *11*, 708–709.
3. Folen, V.J.; Rado, G.T.; Stadler, E.W. Anisotropy of the magnetoelectric effect in Cr_2O_3. *Phys. Rev. Lett.* **1960**, *6*, 607–608. [CrossRef]
4. Martin, T.J.; Anderson, J.C. Antiferromagnetic domain switching in Cr_2O_3. *IEEE Trans. Magn.* **1966**, *MAG 2*, 446–449. [CrossRef]
5. Iyama, A.; Kimura, T. Magnetoelectric hysteresis loops in Cr_2O_3 at room temperature. *Phys. Rev. B* **2013**, *87*, 180408. [CrossRef]
6. Borisov, P.; Ashida, T.; Nozaki, T.; Sahashi, M.; Lederman, D. Magnetoelectric properties of 500-nm Cr_2O_3 films. *Phys. Rev. B* **2016**, *93*, 174415. [CrossRef]
7. Borisov, P.; Hochstrat, A.; Chen, X.; Kleemann, W.; Binek, C. Magnetoelectric switching of exchange bias. *Phys. Rev. Lett.* **2005**, *94*, 117203. [CrossRef]
8. Kosub, T.; Kopte, M.; Radu, F.; Schmidt, O.G.; Makarov, D. All-electric access to the magnetic-field-invariant magnetization of antiferromagnets. *Phys. Rev. Lett.* **2015**, *115*, 097201. [CrossRef] [PubMed]
9. Chen, X.; Hochstrat, A.; Borisov, P.; Kleemann, W. Magnetoelectric exchange bias systems in spintronics. *Appl. Phys. Lett.* **2006**, *89*, 202508. [CrossRef]

10. Toyoki, K.; Shiratsuchi, Y.; Nakamura, T.; Mitsumata, C.; Harimoto, S.; Takechi, Y.; Nishimura, T.; Nomura, H.; Nakatani, R. Equilibrium surface magnetization of α-Cr_2O_3 studied through interfacial chromium magnetization in Co/α-Cr_2O_3 layered structures. *Appl. Phys. Express* **2014**, *7*, 114210. [CrossRef]

11. Ashida, T.; Oida, M.; Shimomura, N.; Nozaki, T.; Shibata, T.; Sahashi, M. Observation of magnetoelectric effect in Cr_2O_3/Co thin film system. *Appl. Phys. Lett.* **2014**, *104*, 152409. [CrossRef]

12. Shiratsuchi, Y.; Watanabe, S.; Yoshida, H.; Kishida, N.; Nakatani, R.; Kotani, Y.; Toyoki, K.; Nakamura, T.; Nakatani, R. Observation of the magnetoelectric reversal process of the antiferromagnetic domain. *Appl. Phys. Lett.* **2018**, *113*, 242404. [CrossRef]

13. Ashida, T.; Oida, M.; Shimomura, M.; Nozaki, T.; Shibata, T.; Sahashi, M. Isothermal electric switching of magnetization in Cr_2O_3/Co thin film system. *Appl. Phys. Lett.* **2015**, *106*, 132407. [CrossRef]

14. Hui, Y.; Lin, W.; Xie, Q.; Chen, S.; Miao, X.; Chen, J. Magnetoelectric effect of epitaxial Cr_2O_3 thin films with a conducting underlayer electrode. *J. Phys. D Appl. Phys.* **2019**, *52*, 24LT03. [CrossRef]

15. Shiratsuchi, Y.; Nakatani, R. Perpendicular exchange bias and magneto-electric control using Cr_2O_3(0001) thin film. *Mater. Trans.* **2016**, *57*, 781–788. [CrossRef]

16. Shiratsuchi, Y.; Toyoki, K.; Tao, Y.; Aono, H.; Nakatani, R. Realization of magnetoelectric effect in 50-nm-thick Cr_2O_3 thin film. *Appl. Phys. Express* **2020**, *13*, 043003. [CrossRef]

17. Wang, J.-L.; Echtenkamp, W.; Mahmood, A.; Binek, C. Voltage controlled magnetism in Cr_2O_3 based all-thin-film systems. *J. Magn. Magn. Mater.* **2019**, *486*, 165262. [CrossRef]

18. Mitsumata, C.; Sakuma, A.; Fukamichi, K.; Tsunoda, M.; Takahashi, M. Critical thickness of antiferromagnetic layer in exchange biasing bilayer system. *J. Phys. Soc. Jpn.* **2008**, *77*, 044602. [CrossRef]

19. Kota, Y.; Imamura, H. Narrowing of antiferromagnetic domain wall in corundum-type Cr_2O_3 by lattice strain. *Appl. Phys. Express* **2017**, *10*, 013002. [CrossRef]

20. Nguyen, T.V.A.; Shiratsuchi, Y.; Yonemura, S.; Shibata, T.; Nakatani, R. Energy condition of isothermal magnetoelectric switching of perpendicular exchange bias in Pt/Co/Au/Cr_2O_3/Pt stacked film. *J. Appl. Phys.* **2018**, *124*, 233902. [CrossRef]

21. Shiratsuchi, Y.; Kuroda, W.; Nguyen, T.V.A.; Kotani, Y.; Toyoki, K.; Nakamura, T.; Suzuki, M.; Nakamura, K.; Nakatani, R. Simultaneous achievement of high perpendicular exchange bias and low coercivity by controlling ferromagnetic/antiferromagnetic interfacial magnetic anisotropy. *J. Appl. Phys.* **2017**, *121*, 073902. [CrossRef]

22. Foner, S. High-field antiferromagnetic resonance in Cr_2O_3. *Phys. Rev.* **1963**, *130*, 183–197. [CrossRef]

23. Shiratsuchi, Y.; Wakatsu, K.; Nakamura, T.; Oikawa, H.; Maenou, S.; Narumi, Y.; Tazoe, K.; Mitsumata, C.; Kinoshita, T.; Nojiri, H.; et al. Isothermal switching of perpendicular exchange bias by pulsed high magnetic field. *Appl. Phys. Lett.* **2012**, *100*, 262413. [CrossRef]

24. Jungwirth, T.; Marti, X.; Wadley, P.; Wunderlich, J. Antiferromagnetic spintronics. *Nat. Nanotechnol.* **2016**, *11*, 231. [CrossRef] [PubMed]

25. Baltz, V.; Manchon, A.; Tsoi, M.; Moriyama, T.; Ono, T.; Tserkovnyak, Y. Antiferromagnetic spintronics. *Rev. Mod. Phys.* **2018**, *90*, 015005. [CrossRef]

26. Jia, J.; Chen, Y.; Wang, B.; Han, B.; Wu, Y.; Wang, Y.; Cao, J. The double-shifted magnetic hysteresis loops and domain structure in perpendicular [Co/Ni]$_N$/IrMn exchange biased systems. *J. Phys. D Appl. Phys.* **2019**, *52*, 065001. [CrossRef]

27. Toyoki, K.; Shiratsuchi, Y.; Kobane, A.; Mitsumata, C.; Kotani, Y.; Nakamura, T.; Nakatani, R. Magnetoelectric switching of perpendicular exchange bias in Pt/Co/Pt/α-Cr_2O_3/Pt stacked films. *Appl. Phys. Lett.* **2015**, *106*, 162404. [CrossRef]

28. Al-Mahdawi, M.; Pati, S.P.; Shiokawa, Y.; Ye, S.; Nozaki, T.; Sahashi, M. Low-energy magnetoelectric control of domain satte in exchange-coupled heterostructures. *Phys. Rev. B* **2017**, *95*, 144423. [CrossRef]

29. Shiratsuchi, Y.; Noutomi, H.; Oikawa, H.; Nakamura, T.; Suzuki, M.; Fujita, T.; Arakawa, K.; Takechi, Y.; Mori, H.; Kinoshita, T.; et al. Detection and in situ switching of unreversed interfacial antiferromagnetic spins in a perpendicular exchange biased system. *Phys. Rev. Lett.* **2012**, *109*, 077202. [CrossRef]

30. Wang, J.; Neaton, J.B.; Zheng, H.; Nagarajan, V.; Ogale, S.B.; Liu, B.; Viehland, D.; Vaithyarathan, V.; Schlom, D.G.; Waghmare, U.V.; et al. Epitaxial $BiFeO_3$ multiferroic thin film heterostructures. *Science* **2003**, *299*, 1719–1722. [CrossRef]

31. Shiratsuchi, Y.; Toyoki, K.; Nakatani, R. Magnetoelectric control of antiferromagnetic domain state in Cr_2O_3 thin film. *J. Phys. Condensed Matter.* **2021**, submitted.

32. Meiklejohn, W.H. Exchange anisotropy—A review. *J. Appl. Phys.* **1962**, *33*, 1328–1335. [CrossRef]

33. Mauri, D.; Siegmann, H.C.; Bagus, P.S.; Kay, E. Simple model for thin ferromagnetic films exchange coupled to an antiferromagnetic substrate. *J. Appl. Phys.* **1987**, *62*, 3047–3049. [CrossRef]

34. Bertotti, G. *Hysteresis in Magnetism*; Academic Press: San Diego, CA, USA, 1998; pp. 17–20.

35. Kosub, T.; Kopte, M.; Hühne, R.; Appel, P.; Shields, B.; Maletinsky, P.; Hübner, R.; Liedke, M.O.; Schimidt, O.G.; Makarov, D. Purely antiferromagnetic magnetoelectric random access memory. *Nat. Commun.* **2016**, *8*, 13985. [CrossRef] [PubMed]

36. Nozaki, T.; Al-Mahdawi, M.; Shiokawa, Y.; Pati, S.P.; Ye, S.; Kotani, Y.; Toyoki, K.; Nakamura, T.; Suzuki, M.; Yonemura, S.; et al. Manipulation of antiferromagnetic spin using tunable parasitic magnetization in magnetoelectric antiferromagnet. *Phys. Status Solidi RRL* **2018**, *12*, 180366. [CrossRef]

magnetochemistry

MDPI

Article

Growth and Characterisation of Antiferromagnetic Ni₂MnAl Heusler Alloy Films

Teodor Huminiuc [1], Oliver Whear [1], Andrew J. Vick [1], David C. Lloyd [2], Gonzalo Vallejo-Fernandez [1], Kevin O'Grady [1] and Atsufumi Hirohata [2,*]

[1] Department of Physics, University of York, Heslington, York YO10 5DD, UK; teodor.huminiuc@gmail.com (T.H.); owhear@gmail.com (O.W.); andrew.vick@stfc.ac.uk (A.J.V.); gonzalo.vallejofernandez@york.ac.uk (G.V.-F.); kevin.ogrady@york.ac.uk (K.O.)
[2] Department of Electronic Engineering, University of York, Heslington, York YO10 5DD, UK; david.lloyd@york.ac.uk
* Correspondence: atsufumi.hirohata@york.ac.uk

Abstract: Recent rapid advancement in antiferromagnetic spintronics paves a new path for efficient computing with THz operation. To date, major studies have been performed with conventional metallic, e.g., Ir-Mn and Pt-Mn, and semiconducting, e.g., CuMnAs, antiferromagnets, which may suffer from their elemental criticality and high resistivity. In order to resolve these obstacles, new antiferromagnetic films are under intense development for device operation above room temperature. Here, we report the structural and magnetic properties of an antiferromagnetic Ni₂MnAl Heusler alloy with and without Fe and Co doping in thin film form, which has significant potential for device applications.

Keywords: antiferromagnets; Heusler alloys; exchange bias; blocking temperature; spintronic devices

Citation: Huminiuc, T.; Whear, O.; Vick, A.J.; Lloyd, D.C.; Vallejo-Fernandez, G.; O'Grady, K.; Hirohata, A. Growth and Characterisation of Antiferromagnetic Ni₂MnAl Heusler Alloy Films. *Magnetochemistry* **2021**, 7, 127. https://doi.org/10.3390/magnetochemistry7090127

Academic Editor: Adam J. Hauser

Received: 4 August 2021
Accepted: 9 September 2021
Published: 13 September 2021

Publisher's Note: MDPI stays neutral with regard to jurisdictional claims in published maps and institutional affiliations.

1. Introduction

Antiferromagnetic spintronics has been attracting a lot of attention due to its potential for THz operation and low power consumption [1]. Currently, antiferromagnetic metals, e.g., IrMn₃ and PtMn₃, and semiconductors, e.g., CuMnAs, have been commonly used [2]. However, these materials are fragile against local compositional changes and may not be suitable for miniaturisation. For further investigation towards device applications, it is critical to develop new antiferromagnetic materials, which are robust against nanofabrication, such as oxidation, atomic mixing, edge roughness and post-annealing.

Antiferromagnetic Heusler alloys are good candidates for applications with precise controllability by potential atomic substitutions. In particular, ternary Heusler alloys, e.g., Pt₂MnGa [3], Ni₂MnAl [4,5] and Mn₂VSi [6], have been reported to exhibit antiferromagnetism as schematically shown in Figure 1. Among them, Ni₂MnAl has been studied and has a lattice constant of 0.5812 nm [5], similar to commonly used seed layers and substrates. However, the Néel temperature has been reported to be between 40 °C [7] and 80 °C [4], which will need to be increased for device applications via atomic substitution as previously reported, e.g., for compensated Mn₂.₄Pt₀.₆Ga ferrimagnet [8]. These studies have been based on epitaxial Ni₂MnAl films [9,10] and bulk samples, but not on polycrystalline films with their suitability for devices, e.g., magnetic sensors and recording. Recent reduction in a device size can utilise a single grain in a polycrystalline film to avoid any electrical and/or magnetic scattering at a grain boundary and/or magnetic domain wall.

In this paper, we report on the optimised growth and annealing conditions for polycrystalline Ni₂MnAl films sputtered at room temperature (RT). The optimisation involves (i) post-annealing between ~400 and 500 °C for up to 2 h and (ii) atomic substitution for crystallisation. Point (ii) is highly significant for the tuning of structural and magnetic properties, such as the Néel temperature, blocking temperature and the lattice constant.

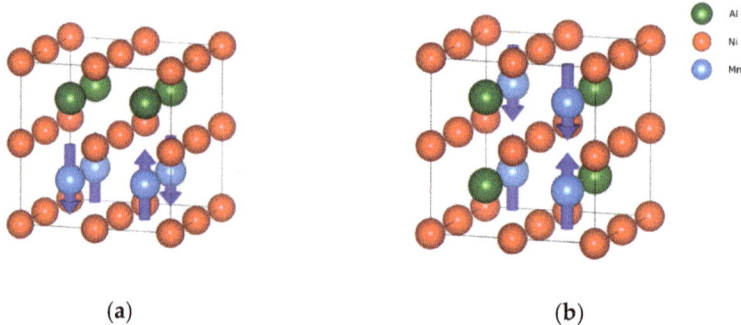

(a)   (b)

Figure 1. Schematic crystal and spin structures of the pseudo-*B*2-(**a**) I and (**b**)-II phases for Ni_2MnAl. Red, blue and green atoms represent Ni, Mn and Al, respectively.

2. Experimental Procedures

100 nm thick Heusler alloy films capped with 2~5 nm of Ru or Ta were deposited onto Si substrates using RF generated remote plasma sputtering. The layers were deposited using a PlasmaQuest high target utilisation sputtering system (HiTUS) with a base pressure of 3×10^{-5} Pa. The plasma was generated by an RF field of 13.56 MHz in an Ar atmosphere of 3×10^{-1} Pa and steered onto the target with a DC bias (V_T) ranging from -250 to -990 V. V_T controlled the deposition rate and the resulting atomic mixing. The typical deposition rate from the stoichiometric Ni_2MnAl target using a bias voltage of 990 V was 0.06 nm/s. The growth rate also controls the grain size. Ni_2MnAl (25)/CoFe (10) and Co_2FeSi (10) (thickness in nm) bilayers were also grown using HiTUS in a similar manner.

The composition of the films was analysed using inductively coupled plasma optical emission spectroscopy (ICP-OES) by InterTek Ltd., confirming near stoichiometry (with some Al deficiency), Ni:Mn:Al = 54.1:28.9:17.0 and 56.1:29.6:14.3 for $V_T = -900$ and -300 V, respectively. The crystalline structures were characterised using out-of-plane (OP) and grazing incidence in-plane (IP) X-ray diffraction (XRD, Rigaku SmartLab) with a Cu $K\alpha$ source and a Ge(220) 2-bounce monochromator. The Heusler-alloy films were annealed in an Ar gas flow (2 L/min.) between 235 and 700 °C for up to 9 h. After each annealing step magnetisation curves were measured using an alternating gradient force magnetometer (AGFM, Princeton Measurements Model 2900) at RT.

3. Results and Discussion

3.1. Ni₂MnAl Films

From XRD measurements, a Ni_2MnAl(220) principal peak was found once samples were annealed at 700 °C as shown in Figure 2. A well-defined (220) peak was developed OP within the first 90 min of annealing. However, after a further one hour of annealing, no further OP crystalline growth was observed as shown in Figure 2b. As similar to $Fe_{2+x}V_yAl$ [11], longer annealing induces phase segregation in the Ni_2MnAl films as confirmed by the appearance of a ferromagnetic phase. The OP lattice constant of Ni_2MnAl was measured to be (0.575 ± 0.001) nm. This is about 1% lower than the theoretically predicted value of 0.5812 nm [5]. There was a transition in the IP (220) peak when the sample was annealed for more than 1.5 h, corresponding to a 1% change in the IP lattice constant from (0.577 ± 0.001) nm to (0.571 ± 0.001) nm, which almost agree with the OP results. The lattice constants of the sputtered polycrystalline Ni_2MnAl films are almost 2% lower than the theoretical value, which may be due to the lower crystallinity and the formation of the segregated phases in the films. This is supported by the absence of additional superlattice (200) or (111) peaks in the XRD patterns. Magnetic measurements confirmed that these films show no magnetic response, indicating that they are in an either antiferromagnetic, paramagnetic or compensated ferrimagnetic state.

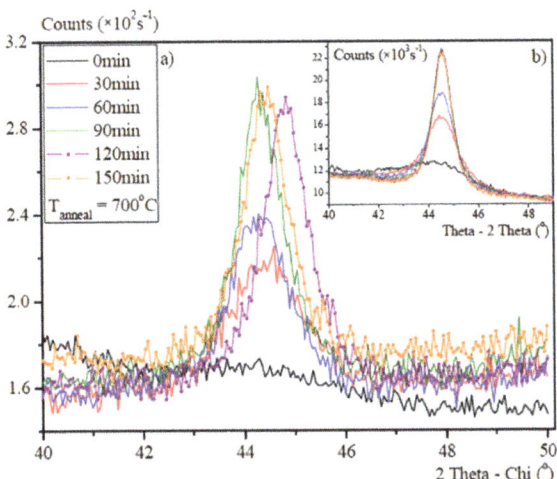

Figure 2. Representative XRD patterns of the Ni$_2$MnAl(220) peaks measured by (**a**) IP and (**b**) OP configurations.

In order to quantify the crystallinity of the Ni$_2$MnAl films measured by XRD, we used a Q factor defined as

$$\text{Q factor} = \frac{\text{Intensity (cps/}^\circ)}{\text{Full width half maximum (FWHM)}} \qquad (1)$$

The (220) peaks were used to quantify the A2 crystalline ordering (atomic mixing between Ni, Mn and Al), i.e., the increase in Q factor represents the increase in A2 ordering, which typically leads to B2 ordering (atomic mixing between Ni and Mn) with a coexisting (200) peak and L2$_1$ perfect ordering with coexisting (200) and (111) peaks [12].

Given the requirements for device fabrication, annealing temperatures above 500 °C are impractical. Therefore, our optimisation criteria were as follows: (i) we identified the post-annealing condition below 500 °C to maximise the Q factor. The increase in the A2 ordering leads to the formation of the pseudo-B2 phase, which is theoretically [5] and experimentally [10] predicted to exhibit antiferromagnetism; (ii) we have substituted some constituent elements to investigate the effect on the crystallisation temperature.

The Ni$_2$MnAl samples were post-annealed for up to 6 h at temperatures in the range 250 to 700 °C. There was no L2$_1$ crystallisation observed in any of the samples post-annealed at a temperature below 700 °C (see Figure 3). The presence of a small number of A2 ordered nanocrystals was indicated in all the as-deposited samples by a weak (220) Ni$_2$MnAl alloy peak at 44.0° with a FWHM of 0.7°. Rocking curves were used to measure the texture of the (220) peak, indicating that the as-deposited grains were aligned within ±15° from the sample surface only for the epitaxial films as reported previously [10]. In a similar Heusler alloy of Fe$_{2+x}$V$_y$Al films [11], a B2 phase has been formed when the Q-factor increases over one order of magnitude as compared with the initial A2 crystallisation occurred. In this study we applied this criterion to identify the phase transformation in Figure 3.

The corresponding exchange bias are shown in Figure 4 for the Ni$_2$MnAl (25)/CoFe (10) and Co$_2$FeSi (10) (thickness in nm) bilayers post-annealed at 250 and 400 °C for 2 h, which were in the A2 and B32a phase, respectively, as shown in Figure 3. These magnetisation curves were measured after field cooling to 100 K under an in-plane magnetic field of 20 kOe. No clear exchange bias effect can be seen within the measured temperature range in Figure 4. However, an increase in the coercivity was observed only for the sample annealed at 400 °C forming the B32a phase. This may be indicative of coupling between layers, suggesting the Ni$_2$MnAl layer is antiferromagnetic but the corresponding Néel (and blocking) temperature is below 100 K possibly due to the coexisting disordered phases. Further increase in the annealing temperature leads to the higher crystallinity of Ni$_2$MnAl as shown in Figure 3 but it induces interfacial mixing in the bilayers causing reduction in the saturation magnetisation. By maintaining the sharp interface between Ni$_2$MnAl and the neighbouring ferromagnetic layer, an exchange bias can be measured at 100 K or above [10].

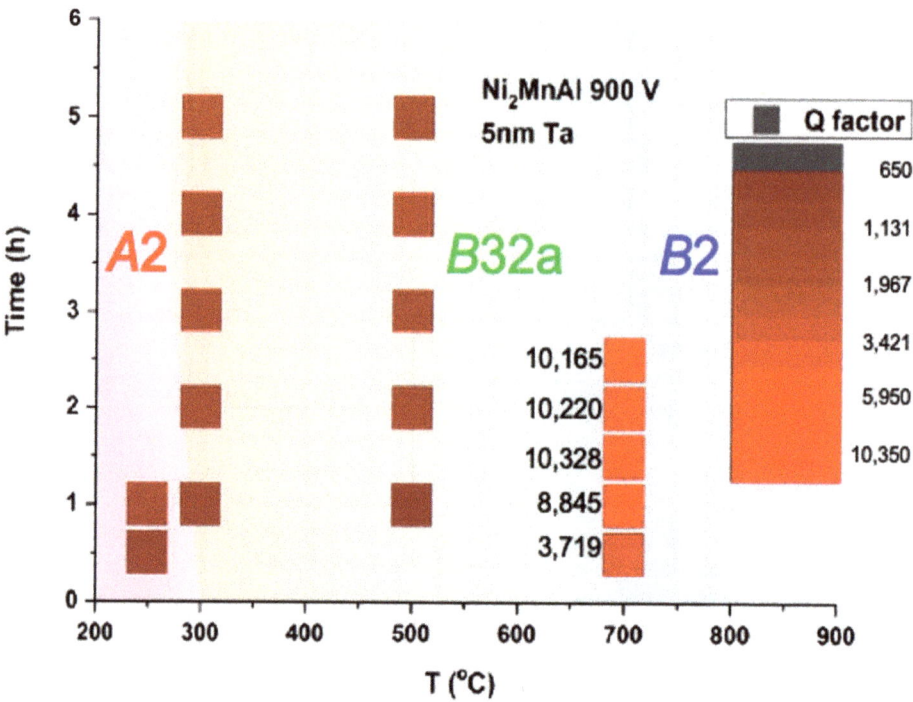

Figure 3. Crystallisation of the Ni$_2$MnAl thin films indicated by Q factors, including annealing conditions which did not lead to crystallisation. The letters shown next to colour scales indicate the corresponding Q-factors.

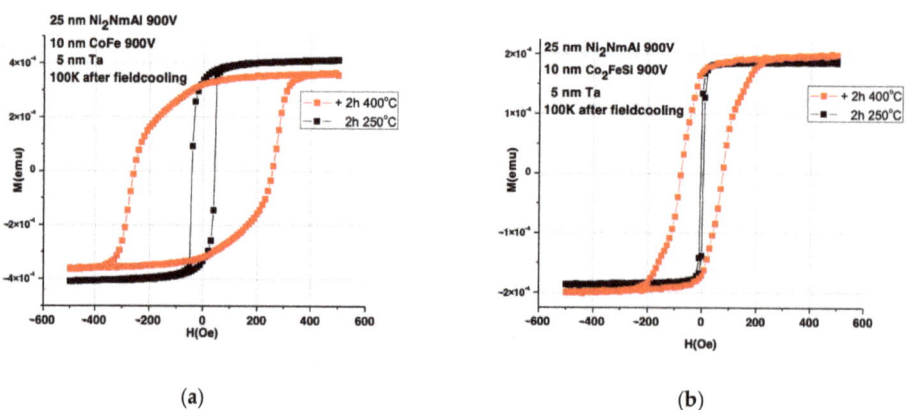

Figure 4. Magnetisation curves of the Ni$_2$MnAl (25)/(**a**) CoFe (10) and (**b**) Co$_2$FeSi (10)/Ta (5) (thickness in nm) bilayer post-annealed at 250 and 400 °C for 2 h under the field applications in the plane.

3.2. Ni₂MnAl Films with Atomic Substitutions by Al

As the sputtered Ni_2MnAl films are ~10% Al deficient, one 6 mm diameter Al peg was added to a pre-drilled hole in the sputtering target, achieving Ni:Mn:Al = 49:27:24. With a target utilisation of over 90%, this technique allows for the variation of the film composition in the HiTUS. Figure 5 shows the Q factors for the Al-doped Ni_2MnAl post-annealed at 500 °C, which was found to be the maximum annealing temperature to avoid Al segregation. Similar to the films without Al doping, Ni_2MnAl crystallises at a relatively high temperature. The Al-doped Ni_2MnAl films have therefore the optimum *A2* crystal structure after 2~4 h of annealing at 500 °C. The corresponding lattice constants were found to decrease from 0.585 to 0.575 nm after 1 and 4 h of annealing, respectively. This confirms that the lattice constants are similar to those without Al doping except for the sample annealed for 1 h. This agrees with the increase in the Q factors shown in Figure 5, indicating the effect of the Al doping on the optimised crystallisation temperature and crystallinity of Ni_2MnAl is negligible.

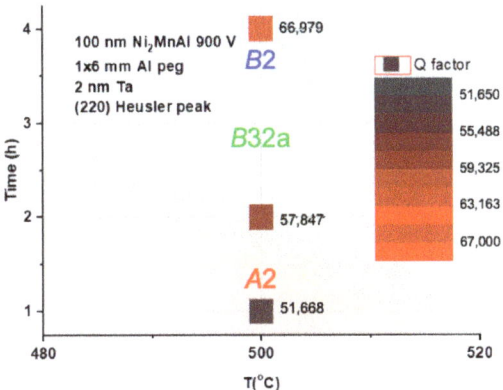

Figure 5. Crystallisation of the ordered Ni_2MnAl thin films doped with a 6 mm Al peg indicated by Q factors.

3.3. Ni₂MnAl Films with Atomic Substitutions by Fe

In order to decrease the Ni_2MnAl crystallisation temperature, an attempt to substitute Ni in the lattice by Fe was made because its covalent radius is greater than that of Ni. This results in the lattice being strained and this may decrease the manganese–manganese nearest neighbour distance to induce antiferromagnetic coupling. However, no crystallisation was found. $Fe_{2+x}V_yAl$ has been studied previously [11], showing that the ordering temperature is lower than that of the Ni_2MnAl thin films. Therefore, it is predicted that doping with Fe can lower the ordering temperature and induce antiferromagnetic ordering in the Ni_2MnAl thin films.

The Ni_2MnAl sputtering targets were doped with two or four Fe pegs with a diameter of 2 mm, effectively making a $Ni_{2-x}Fe_xMnAl$ target. The sputtered Ni_2MnAl films doped with Fe were found to exhibit ferromagnetic behaviour even in the as-deposited state. Almost negligible magnetic moments of (1.5 ± 0.1) μemu, approximately 3×10^{-2} emu/cm³, was measured for the samples grown from the Ni_2MnAl sputtering target doped with two Fe pegs and (2.7 ± 0.1) μemu with four Fe pegs. This small moment was probably induced by Fe segregation. The films were then post-annealed for up to three hours at temperatures between 400 and 500 °C. However, no crystallisation of the Ni_2MnAl Heusler alloy was detected but the average saturation magnetisation increased to (2.0 ± 0.1) μemu and (3.4 ± 0.1) μemu after annealing at 500 °C for 3 h for the films doped with two and four Fe pegs, respectively. This suggests that the Fe segregation is promoted by post-annealing.

3.4. Ni$_2$MnAl Films with Atomic Substitutions by Co

Co-doping was also tested because its covalent radius is greater than that of Ni but is smaller than that of Fe. This may also decrease the manganese–manganese nearest neighbour distance to induce antiferromagnetic coupling as predicted for the case of the Fe doping. The corresponding lattice constants for one Co peg were found to decrease from 0.584 to 0.580 nm after 1 and 5 h of annealing, respectively. Co$_2$MnAl has hence been reported to have an ordering temperature lower than Ni$_2$MnAl [13,14], suggesting that the Co doping may also lower the ordering temperature and promote antiferromagnetic ordering.

The Ni$_{2-x}$Co$_x$MnAl sputtering targets were used adding one or two Co pegs with a diameter of 6 mm onto the Ni$_2$MnAl target, achieving Ni:Co:Mn:Al = 56:9:20:15 and 51:15:20:14 for one and two Co pegs, respectively. The sputtered films exhibited paramagnetic behaviour in the as-deposited state. The films successfully crystallised after post-annealing at temperatures between 400 and 500 °C for two hours as shown in Figure 6. Both samples exhibited the *A*2 ordering as identified by the (220) peak at 43°. No clear (200) peak was visible at 37.5° which again suggests the absence of *B*2 ordering.

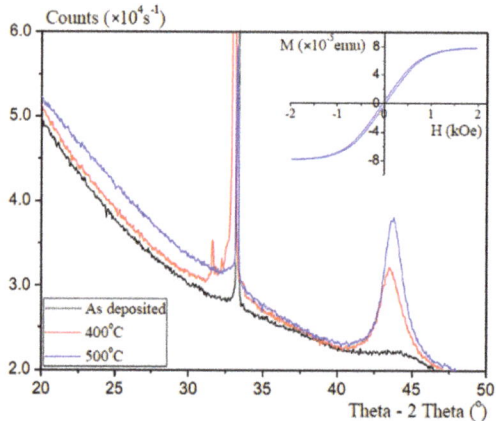

Figure 6. Representative XRD patterns of the Ni$_2$MnAl thin films doped with one Co peg as deposited and after two hours of annealing at 400 and 500 °C. The inset shows the corresponding magnetisation loop of the latter film. Note that the silicon substrate (200) peak and the fringes are visible at 33.2°.

As shown in Figure 7, the crystallisation also occurs at 300 °C, which is 150 °C lower than that for films without Co doping, which is favourable for device applications. The corresponding lattice constants were measured to be 0.8~2.9% larger than those for films without Co doping.

The crystallinity was also found to improve with increasing Co doping as shown in Figure 7. The magnetic response of the samples changed to ferromagnetic after post-annealing at 500 °C as shown in the inset of Figure 6, showing a magnetic moment of (76.2 ± 0.1) μemu, approximately 1.5 emu/cm^3, which is larger than that of the Fe substituted films. A possible cause for the ferromagnetism is the formation of Co$_2$MnAl grains because the films were annealed at the ordering temperature [15]. No further crystalline ordering was observed with increasing annealing and the corresponding magnetic properties were unchanged. This observation supports the hypothesis of the formation of Co$_2$MnAl as the cobalt content can be segregated from the Ni$_2$MnAl matrix, thus preventing further significant formation of grains. To eliminate such segregated grains, uniform compositional distributions within a film may be required, which is difficult to achieve using the doping method of the sputtering target used in this study. Even so, our study suggests a significant potential of Ni$_{2-x}$Co$_x$MnAl for robust antiferromagnetism at room temperature.

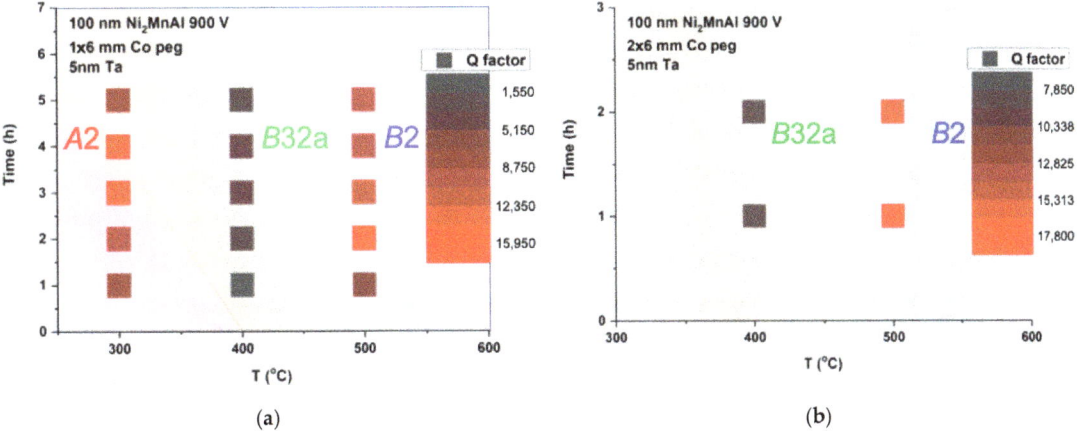

Figure 7. Crystallisation of the Ni$_2$MnAl thin films doped with (**a**) one and (**b**) 6 mm Co pegs indicated by Q factors.

4. Summary

We have grown and characterised a series of Ni$_2$MnAl Heusler alloy films. The films with one Co peg post-annealed at 500 °C for 2 h were found to show the highest crystalline ordering within the *A*2 phase but exhibited weak magnetic moments. For the demonstration of room-temperature antiferromagnetism, Fe and Co have been used to partially substitute for Ni. The Fe substitution showed an increase in the magnetic moment with increasing Fe content, which may be due to Fe segregation. On the other hand, Co substitution can effectively reduce the crystallisation temperature down to 300 °C but the corresponding magnetisation measurements proved that the *B*2-ordered (Ni,Co)$_2$MnAl films showed ferromagnetic Co$_2$MnAl segregation as well. The segregated phases need to be eliminated by further compositional optimisation to achieve the stoichiometry while maintaining the reduced crystallisation in the pseudo-*B*2 phase temperature for antiferromagnetic spintronics.

Author Contributions: All authors contributed to write this article. T.H. and O.W. grew and characterised the antiferromagnetic Heusler alloy films. A.J.V. and D.C.L. maintained and assisted with structural and magnetic characterisation. G.V.-F. and K.O. helped the analysis of exchange bias. A.H. developed the imaging method and analysed the data. A.H. conceived the experiment and analysed the results. All authors have read and agreed to the published version of the manuscript.

Funding: This work is partially funded by EU-FP7 HARFIR (NMP3-SL-2013-604398) and Physical Sciences Research Council (EPSRC) (EP/M02458X/1, EP/M02458X/1 and EP/V007211/1) and Japan Science and Technology Agency (JST) Core Research for Evolutional Science and Technology (CREST) (JPMJCR17J5).

Institutional Review Board Statement: Not applicable.

Informed Consent Statement: Not applicable.

Data Availability Statement: Data is contained within the article and available on request with following the guideline set by the University of York (UK).

Conflicts of Interest: The authors declare no conflict of interest.

References

1. Jungwirth, T.; Marti, X.; Wadley, P.; Wunderlich, J. Antiferroagnetic spintronics. *Nat. Nanotech.* **2016**, *11*, 231–241. [CrossRef] [PubMed]
2. Hirohata, A.; Huminiuc, T.; Sinclair, J.; Wu, H.; Samiepour, M.; Vallejo-Fernandez, G.; O'Grady, K.; Balluf, J.; Meinert, M.; Reiss, G.; et al. Development of antiferromagnetic Heusler alloys for the replacement of iridium as a critically raw material. *J. Phys. D Appl. Phys.* **2017**, *50*, 443001. [CrossRef]
3. Singh, S.; D'Souza, W.; Suard, E.; Chapon, L.; Senyshyn, A.; Petricek, V.; Skourski, Y.; Nicklas, M.; Felser, C.; Chadov, S. Room-temperature tetragonal non-collinear Heusler antiferromagnet Pt_2MnGa. *Nat. Commun.* **2016**, *7*, 1–6. [CrossRef] [PubMed]
4. Acet, M.; Duman, E.; Wassermann, E.F. Coexisting ferro- and antiferromagnetism in Ni_2MnAl Heusler alloys. *J. Appl. Phys.* **2002**, *92*, 3867–3871. [CrossRef]
5. Galanakis, I.; Şaşıoğlu, E. Structural-induced antiferromagnetism in Mn-based full Heusler alloys: The case of Ni_2MnAl. *Appl. Phys. Lett.* **2011**, *98*, 102514. [CrossRef]
6. Wu, H.; Vallejo-Fernandez, G.; Hirohata, A. Magnetic and structural properties of antiferromagnetic Mn_2VSi alloy films grown at elevated temperatures. *J. Phys. D Appl. Phys.* **2017**, *50*, 375001. [CrossRef]
7. Singh, D.J.; Mazin, I. Electronic structure, local moments, and transport in Fe_2VAl. *Phys. Rev. B* **1998**, *57*, 14352. [CrossRef]
8. Nayak, A.K.; Nicklas, M.; Chadov, S.; Khuntia, P.; Shekhar, C.; Kalache, A.; Baenitz, M.; Skourski, Y.; Guduru, V.K.; Puri, A.; et al. Design of compensated ferrimagnetic Heusler alloys for giant tunable exchange bias. *Nat. Mater.* **2015**, *14*, 679–684. [CrossRef] [PubMed]
9. Dong, X.Y.; Dong, J.W.; Xie, J.Q.; Shih, T.C.; McKernan, S.; Leighton, C.; Palmstrøm, C.J. Growth temperature controlled magnetism in molecular beam epitaxially grown Ni_2MnAl Heusler alloy. *J. Cryst. Growth* **2003**, *254*, 384–389. [CrossRef]
10. Tsuchiya, T.; Kubota, T.; Sugiyama, T.; Huminiuc, T.; Hirohata, A.; Takanashi, K. Exchange bias effects in Heusler alloy Ni_2MnAl/Fe bilayers. *J. Phys. D Appl. Phys.* **2016**, *49*, 235001. [CrossRef]
11. Huminiuc, T.; Whear, O.; Takahashi, T.; Kim, J.-Y.; Vick, A.; Vallejo-Fernandez, G.; O'Grady, K.; Hirohata, A. Growth and characterisation of ferromagnetic and antiferromagnetic $Fe_{2+x}V_yAl$ Heusler alloy films. *J. Phys. D Appl. Phys.* **2018**, *51*, 325003. [CrossRef]
12. Elphick, K.; Frost, W.; Samiepour, M.; Kubota, T.; Takanashi, K.; Sukegawa, H.; Mitani, S.; Hirohata, A. Heusler alloys for spintronic devices: Review on recent development and future perspectives. *Sci. Technol. Adv. Mater.* **2021**, *22*, 235–271. [CrossRef] [PubMed]
13. Kanomata, T.; Kikuchi, M.; Yamauchi, H. Magnetic properties of Heusler alloys Ru_2MnZ (Z = Si, Ge, Sn and Sb). *J. Alloys Compd.* **2006**, *414*, 1–7. [CrossRef]
14. Ishida, S.; Kashiwagi, S.; Fujii, S.; Asano, S. Magnetic and half-metallic properties of new Heusler alloys Ru_2MnZ (Z = Si, Ge, Sn and Sb). *Phys. B* **1995**, *210*, 140–148. [CrossRef]
15. Sagar, J.; Fleet, L.R.; Walsh, M.; Lari, L.; Boyes, E.D.; Whear, O.; Huminiuc, T.; Vick, A.; Hirohata, A. Over 50% reduction in the formation energy of Co-based Heusler alloy films by two-dimensional crystallisation. *Appl. Phys. Lett.* **2014**, *105*, 032401. [CrossRef]

MDPI

Article

Large Perpendicular Exchange Energy in $Tb_xCo_{100-x}/Cu(t)/[Co/Pt]_2$ Heterostructures

Sina Ranjbar *, Satoshi Sumi, Kenji Tanabe and Hiroyuki Awano

Toyota Technological Institute, Nagoya 468-8511, Japan; sumi@toyota-ti.ac.jp (S.S.); tanabe@toyota-ti.ac.jp (K.T.); awano@toyota-ti.ac.jp (H.A.)
* Correspondence: sina.ranjbar@toyota-ti.ac.jp

Abstract: In order to realize a perpendicular exchange bias for applications, a robust and tunable exchange bias is required for spintronic applications. Here, we show the perpendicular exchange energy (PEE) in the $Tb_xCo_{100-x}/Cu/[Co/Pt]_2$ heterostructures. The structure consists of amorphous ferrimagnetic Tb–Co alloy films and ferromagnetic Co/Pt multilayers. The dependence of the PEE on the interlayer thickness of Cu and the composition of Tb–Co were analyzed. We demonstrate that the PEE can be controlled by changing the Cu interlayer thickness of $0.2 < t_{Cu} < 0.3$ (nm). We found that PEE reaches a maximum value ($\sigma_{Pw} = 1$ erg/cm^2) at around x = 24%. We, therefore, realize the mechanism of PEE in the $Tb_xCo_{100-x}/Cu/[Co/Pt]_2$ heterostructures. We observe two competing mechanisms—one leading to an increase and the other to a decrease—which corresponds to the effect of Tb content on saturation magnetization and the coercivity of heterostructures. Sequentially, our findings show possibilities for both pinned layers in spintronics and memory device applications by producing large PEE and controlled PEE by Cu thickness, based on $Tb_xCo_{100-x}/Cu/[Co/Pt]_2$ heterostructures.

Keywords: perpendicular magnetic anisotropy; ferrimagnet; perpendicular exchange bias; amorphous thin films; spintronic applications

Citation: Ranjbar, S.; Sumi, S.; Tanabe, K.; Awano, H. Large Perpendicular Exchange Energy in $Tb_xCo_{100-x}/Cu/[Co/Pt]_2$ Heterostructures. *Magnetochemistry* **2021**, 7, 141. https://doi.org/10.3390/magnetochemistry7110141

Academic Editor: Atsufumi Hirohata

Received: 28 September 2021
Accepted: 20 October 2021
Published: 25 October 2021

Publisher's Note: MDPI stays neutral with regard to jurisdictional claims in published maps and institutional affiliations.

1. Introduction

The exchange bias (EB) phenomenon was discovered more than half a century ago by Meiklejohn and Bean [1]. EB can be observed through the exchange coupling between ferromagnet (FM)/antiferromagnetic (AFM) layers at the interface [2,3]. Utilizing a large perpendicular EB field as pinned layers in giant-magnetoresistive (GMR) devices, hard-disk drives (HDDs), magnetic random-access memory (MRAM) technologies, and magnetic tunnel junctions (MTJs) have been the subjects of intense attraction because of their potential in spintronic applications [4–9]. On the other hand, in memory device applications, controlling the perpendicular exchange energy (PEE) is a crucial factor [10–13].

The exchange anisotropy energy is generally revealed by the exchange energy, σ_{Pw}, which is the stabilizing energy per unit area of the FM/AFM or ferrimagnet (FIM)/FM interfaces, $H_{ex} = \frac{J_k}{M_s t} = \frac{\sigma_{Pw}}{2M_s t}$, where M_s and t_{FM} are the saturation magnetization and thickness of the FM layer, respectively [14,15]. However, typical AFM/FM systems indicate a limitation in attaining large EB fields (usually below 1 KOe) [16–18], which correlates to challenges in fabricating fine AFM crystals, controlling the AFM domain state, and uncompensated spin moments at the interface [16,17,19]. Thus, it seems that these cannot provide the reasonable necessities for future spintronic applications. Aside from FM/AFM systems, exchange bias also exists in ferrimagnet FIM/FIM [20] and the ferromagnet FIM/FM bilayer [21]. Amorphous rare earth-transition metal (RE-TM) multilayers exhibit strong perpendicular magnetic anisotropy (PMA) and robust coupling interactions at the interface [22,23].

In amorphous ferrimagnetic (FI) rare earth-transition metal (RE-TM) alloy films, there are two kinds of pair interactions, the antiparallel exchange between the RE-TM moments and the parallel exchange of the TM moments themselves; both interactions can provide a sufficiently strong interlayer coupling with the adjacent FM layer, and can hence provide a higher EB field—even for fully compensated interfaces [24,25].

Few kinds of research regarding the FIM/FIM, FM/FIM composite structures have been performed. Due to their potential for spintronic applications, more investigation is required to understand the mechanisms of these systems [20,26,27].

In this paper, we investigate the perpendicular exchange energy (PEE) between Tb–Co alloy films and Co/Pt multilayers by varying the Tb content of the Tb–Co layer. We observe that both EB and PEE can be tuned by introducing a Cu spacer layer. The PEE attains its maximum $\sigma_{Pw} = 1$ erg/cm^2 at x = 24. We also describe the reason for high PEE at x = 24 in the Tb$_x$Co$_{100-x}$(20)/Cu(t)/[Co(0.4)/Pt(2)]$_2$ heterostructures. Our findings prove that Tb$_x$Co$_{100-x}$(20)/Cu(t)/[Co(0.4)/Pt(2)]$_2$ heterostructures are efficient for spintronic applications, and they have advantages for manipulation by adjusting the Tb concentration.

2. Experimental Method

Figure 1a shows a schematic illustration of our SiO$_2$/Pt(5)/Tb$_x$Co$_{100-x}$(20)/Cu(t)/[Co(0.4)/Pt(2)]$_2$ (thicknesses in nm) (20 < x < 41) samples, where the numbers in parentheses indicate the thickness in nanometers. The samples are deposited, using an ultrahigh-vacuum magnetron-sputtering system, onto a silicon substrate at room temperature. Co/Pt multilayers were sputtered, through dc sputtering, using two separate targets of platinum and cobalt at an argon-gas pressure level of 0.2 pascal. Tb$_x$Co$_{100-x}$ films were prepared through co-sputtering using two separate targets of terbium and cobalt. The sample holder rotates during the deposition to ensure a uniform film composition. The thickness of the Tb$_x$Co$_{100-x}$ layer is fixed at 20 nm. The composition of the films was measured using energy-dispersive X-ray analysis (EDX). The magnetic properties were measured at room temperature using the polar-magneto-optical Kerr effect (PMOKE) and a vibrating sample magnetometer (VSM).

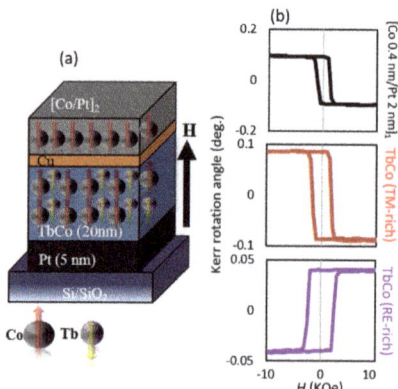

Figure 1. (**a**) Schematic view of the sample layer structure; (**b**) out-of-plane magneto-optic Kerr effect.

3. Results and Discussion

We first determined the magnetic properties of Tb$_x$Co$_{100-x}$(20)/Cu(0.2)/[Co(0.4)/Pt(2)]$_2$ systems. Figure 1b shows the hysteresis loops were measured using the polar-magneto-optical Kerr effect (PMOKE) at room temperature for three single-layer samples, a [Co/Pt]$_2$ multilayer film, and Tb–Co films. All samples show the easy axis perpendicular to the film plane. The polarity of the Kerr rotation (θ_K) signals switches, which is consistent with a transition from being Co dominated to being Tb dominated in the magnetic moment.

(MOKE) signal for a single layer of [Co/Pt] and Tb–Co. The MOKE measurement wavelength is 690 nm for visible light. At 690 nm, only the magneto-optical Kerr effect of Co can be measured. This magneto-optical hysteresis of Co shows negative polarity, as shown in Co/Pt in Figure 1b. Similarly, in the TM-rich Tb–Co single-layer sample in

Figure 1b, since the Co in the Tb–Co layer is aligned in the magnetic field direction, the polarity of the hysteresis is negative, as in Co/Pt. On the other hand, in the RE-rich Tb–Co single-layer sample in Figure 1b, the polarity of hysteresis is negative because the Co is aligned in the opposite direction to the magnetic field.

Figure 2a shows the out-of-plane magnetic hysteresis loops were measured using the VSM at room temperature for $Tb_xCo_{100-x}(20)/Cu(0.2)/[Co(0.4)/Pt(2)]_2$ systems with different Tb concentrations. Here, we used the VSM to find the saturation magnetization of the sample, which is summarized in Figure 2b. Figure 2a shows a two-step switching loop, where the first and second loop switches at the low and high magnetic fields correspond to the Tb–Co and [Co/Pt] multilayers, respectively.

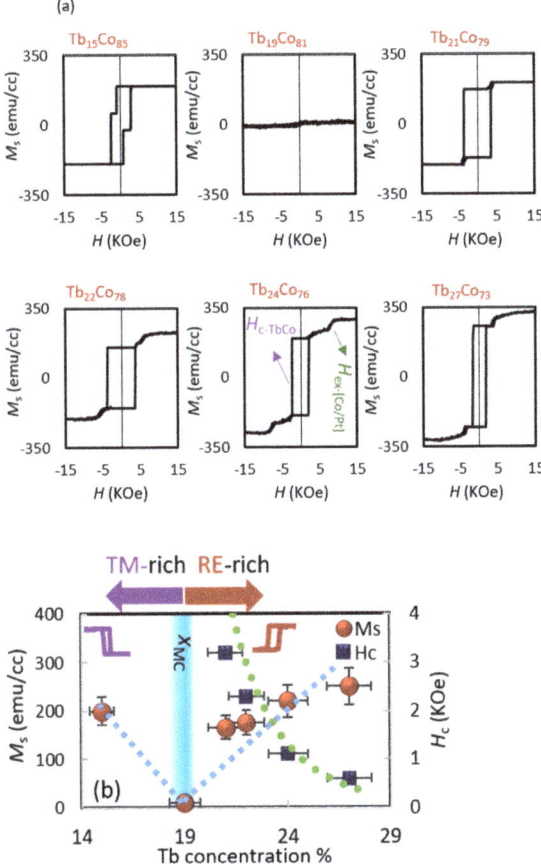

Figure 2. (a) The out-of-plane M–H loop for $Tb_xCo_{100-x}/Cu(0.2)/[Co(0.4)/Pt(2)]_2$ heterostructures, and (b) M_s and H_c as a function of Tb concentration.

Figure 2b shows the coercive fields (H_c) and the saturation magnetizations (M_s) of $Tb_xCo_{100-x}(20)/Cu(0.2)/[Co(0.4)/Pt(2)]_2$ films at different compositions. From the magnetization curve of the $SiO_2/Pt(5)/Tb_xCo_{100-x}(20)/Cu(0.2)/[Co(0.4)/Pt(2)]_2$ heterostructures, it is seen that the saturation magnetization the M_s of Tb_xCo_{100-x} reaches its magnetization compensation composition point at $x_C \sim 19$. While the M_s is at its minimum, the coercive fields reach their maximum at the magnetic compensation composition [28–31]. On the other hand, the perpendicular magnetic anisotropy (K_u) can be calculated using the following equation:

$$K_u \approx \alpha\, M_s\, H_c \tag{1}$$

where the α is constant for all samples, since all heterostructures are prepared under the same conditions. By increasing the Tb concentration the K_u value decreases for all samples, which is in good agreement with the previous report [26,32,33].

To realize the insertion layer effect on H_{ex} and the σ_{Pw}, Figure 3a shows the out-of-plane minor loops (+15 KOe = ⇒ 0 Oe = ⇒ + 15 KOe) for the $Tb_{21}Co_{79}/Cu(t_{Cu})/[Co(0.4)/Pt(2)]_2$ heterostructures ($0.2 < t_{Cu} < 1$ nm). This hysteresis loop is shifted away from the zero-field axis to $H = +H_{ex}$, and the width of the loop is $2\,H_c$, where H_c is the coercive field of a $[Co/Pt]_2$ layer. The hysteresis curve on the high magnetic field side, shown in Figure 2, shows the magnetization reversal of the Co/Pt layer, as evidenced by the negative polarity of the minor loop, as shown in Figure 3a. If this was the result of RE-rich Tb–Co, the polarity would be positive.

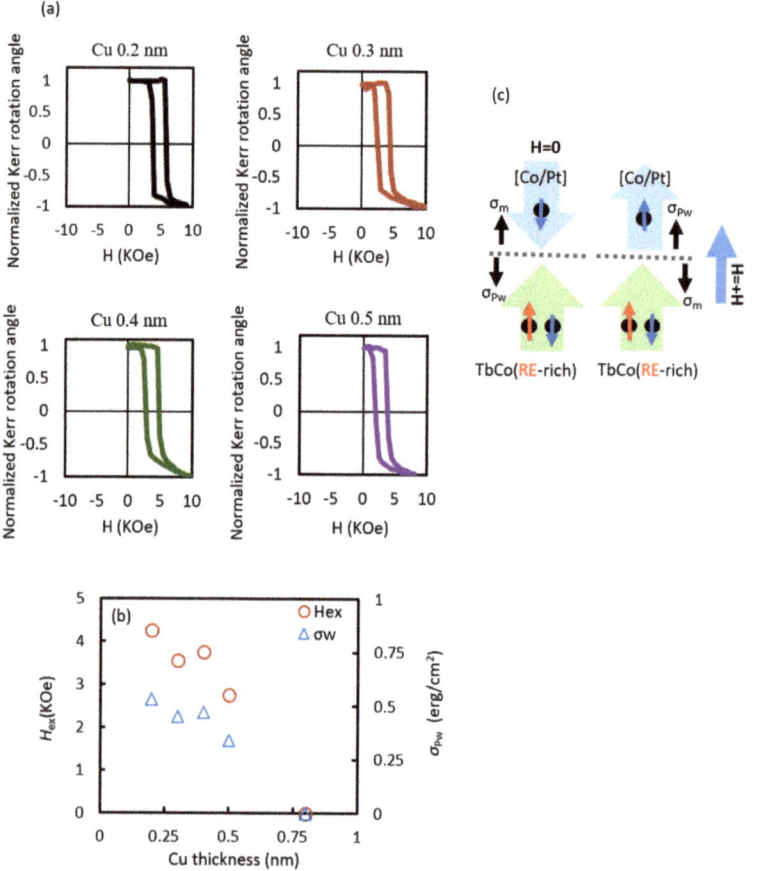

Figure 3. (**a**) Minor loops of the $Tb_{21}Co_{79}/Cu(t_{Cu})/[Co(0.4)/Pt(2)]_2$ heterostructures with various insertion layer thickness; (**b**) changes in the unidirectional anisotropy constant, σ_{Pw}, and exchange anisotropy, H_{ex}, as a function of Cu thickness; (**c**) schematic illustration of the magnetic configuration at the interface of the $Tb_xCo_{100-x}/Cu(0.2)/[Co(0.4)/Pt(2)]_2$ heterostructures.

The Cu interlayer, with a thickness of t_{Cu} = 0.2–1 nm, was employed to tune the perpendicular exchange energy (PEE), σ_{Pw}. The optimization of the insertion layer thickness is necessary to control the PEE value because the removal of the Cu layer causes the exchange coupling to become very large; therefore, it is impractical for the observation of the shift of the hysteresis loop. Ideally, the effect of the Cu insertion causes the intermixing of Tb–Co–Co/Pt that appears during the sputtering process to decrease. Hence, inserting thin Cu layers allows for the improvement of the interface, thus enhancing the effective anisotropy of the stack [34]. Figure 3b shows that by increasing the t_{Cu} over 0.5 nm, the H_{ex} and PEE σ_{Pw} monotonically decrease, which is in agreement with previous reports [35,36]; thus, the optimum H_{ex} and σ_{Pw} values were 4.25 kOe and 0.54 erg/cm^2 at t_{Cu} = 0.2 nm, respectively. As a result, it is clearly demonstrated that the PEE can be controlled by Cu thickness, which indicates additional suitability for memory device applications [10,11].

Figure 3c shows a schematic illustration of the magnetic configuration at the interface of the Tb$_x$Co$_{100-x}$/Cu(0.2)/[Co(0.4)/Pt(2)]$_2$ heterostructures. As shown by the zero magnetic field in the model diagram, the magnetization of the Co/Pt layer and the net magnetization of the Tb–Co layer are opposite from one another; therefore, the static magnetic energy at the interface increases. However, since the Co in the Co/Pt layer and the Co in the Tb–Co layer are both oriented in the same direction, the interfacial domain wall energy is low. On the other hand, in the high magnetic field in the model diagram, the static magnetic energy at the interface decreases because the Co/Pt layer is inverted, and conversely, the interfacial domain wall energy increases. In general, when the Co/Pt layer and the Tb–Co layer are directly heterojunctioned, the interfacial domain wall energy is overwhelmingly larger than the interfacial static magnetic energy; thus, it can only be used only for a large exchange bias application. However, it was found that by inserting a small amount of Cu at these interfaces, the interfacial domain wall energy can be controlled and reduced to the desired value. As a result, it can be applied to memory applications that also utilize interfacial static magnetic energy.

To define the EB for each of the heterostructures, the out-of-plane minor loops were measured as shown in Figure 4a. Figure 4b summarized σ_{Pw} and the magnitude of the H_{ex} field, respectively, as a function of the Tb content in the Tb$_x$Co$_{100-x}$/Cu(0.2)/[Co(0.4)/Pt(2)]$_2$ heterostructures at room temperature. Between 23 and 25 at.% Tb, the PEE seems to reach its maximal value; toward lower and higher amounts of Tb, some reduction appears. In the first region, both the H_{ex} and σ_{Pw} values increase when the Tb is among $19 < x < 24$ atomic percent. Contrarily, in the second region—by increasing the Tb content from $x = 24$ [31]—both the H_{ex} and σ_{Pw} values decrease and become zero at around $x = 30$.

In previous studies, it was reported that the EB field reaches its maximum at the compensation point since the compensated sublattices of the FIM film hold no frustrated bonds at the interface to the FM layer [25,37]. However, this behavior cannot be explained directly in our system.

Firstly, this behavior can be explained by taking into account the variation in the exchange energies of the Co–Co pair with changing RE content [38]. Accordingly, the exchange coupling between the Co–Co pairs is the strongest when compared to other pairings ($J_{Co-Co} > J_{Co-Tb} > J_{Co-Pt}$). The J_{Co-Co} is present at the interface between Tb–Co and [Co/Pt]; therefore, a maximum PEE appears at a lower Tb concentration. Increasing the Tb concentration can decrease the number of Co atoms at the interface, and occasionally it can cause the reduction in the exchange coupling between Co–Co pairs at the interface [21,39].

Moreover, by increasing the Tb content, H_c decreases while M_s increases, which is shown in Figure 2b. Therefore, the variation in the σ_{Pw} value in the reduction in H_c is likely connected with a smaller perpendicular exchange energy, σ_{Pw}, in the ferrimagnet, which should result in smaller perpendicular exchange energy.

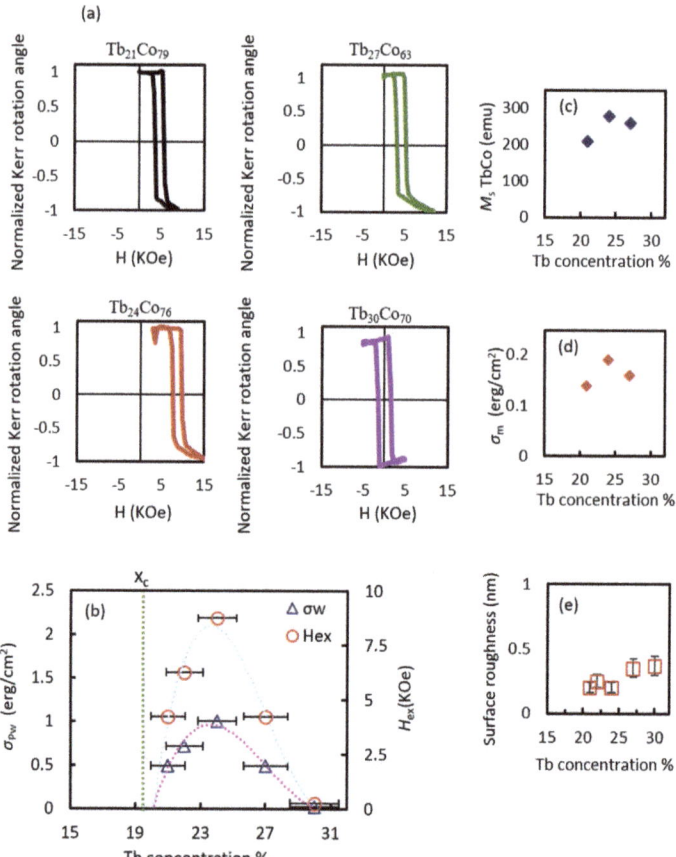

Figure 4. (**a**) Minor loops of the $Tb_xCo_{100-x}/Cu(0.2)/[Co(0.4)/Pt(2)]_2$ heterostructures; (**b**) changes in perpendicular exchange energy, σ_{Pw}, and exchange anisotropy, H_{ex}, as a function of Tb composition; (**c**) Tb–Co magnetization as a function of the Tb content; (**d**) change in magnetostatic energy as a function of Tb composition; (**e**) surface roughness plotted as a function of Tb concentration.

To clarify the variation in σ_{Pw}, the exchange energy can be calculated by the relation in Equation (2). Hence, the total magnetic energy at the interface can be explained by the following equation:

$$E = \sigma_{Pw} + \sigma_m \tag{2}$$

Here, $\sigma_{Pw} = \sigma_{iw} + \sigma_A$, where the first term is the interfacial domain wall energy generated between the Co/Pt layer and the Tb–Co layer; the second term, σ_m, is the static magnetic energy generated between the Co/Pt layer and the Tb–Co layer; and the third term, σ_A, is the magnetic anisotropy generated between the Co/Pt layer and the Tb–Co layer. The interfacial domain wall energy obtained from the inverting magnetic field H_{EX} of the Co/Pt layer is σ_{Pw}.

The anisotropic energy is very small because the Cu intermediate layer greatly attenuates the exchange force between the Co/Pt layer and the Tb–Co layer. Therefore, this σA term can be negligible [40]. To realize the mechanism of the PEE at the interface, we extracted the magnetization information of the samples in Table 1 and summarized them in Figure 4c.

Table 1. Summarized magnetic properties of Tb–Co/[Co/Pt]$_2$ multilayers.

FIM Composition	Ms–Tb–Co (emu/cm^3)	Ms-[Co/Pt] (emu/cm^3)
Tb21Co79	210	1580
Tb24Co76	280	1430
Tb27Co73	260	1550

Figure 4d shows the magnetostatic energy of the Tb$_x$Co$_{100-x}$. It is seen that the magnetostatic energy shows the same curvature, which is in good agreement with Figure 4c. Furthermore, the value of magnetostatic energy is considerable and plays an important role in this system [41].

Since EB is an interfacial phenomenon, the surface roughness may affect the magnitude of the exchange bias [19,42,43]. Hence, the effect of the Tb content on the surface morphology of Tb$_x$Co$_{100-x}$/Cu(0.2)/[Co(0.4)/Pt(2)]$_2$ heterostructures was investigated using AFM. Figure 4e shows a flat surface for Tb$_{24}$Co$_{76}$ in comparison to the other composition. Surface roughness (R$_a$) for Tb–Co increases with increasing Tb concentration. Therefore, the variation in the exchange bias field as a function of the Tb composition might be related to the interface roughness induced by changes in growth conditions, depending on the Tb content of the alloy.

For the Tb$_x$Co$_{100-x}$ samples, both H_{ex} and σ_{Pw} reach a maximum value of H_{ex} = 8.75 kOe and σ_{Pw} = 1 (erg/cm^2) at x = 24, which are significantly larger than what was observed in the ordinary AFM/FM and FM/FM systems [17,27,36,44,45].

4. Summary

In summary, we have systematically investigated the perpendicular exchange bias and perpendicular exchange energy (PEE) σ_{Pw} of Tb$_x$Co$_{100-x}$/Cu(t_{Cu})/[Co(0.4)/Pt(2)]$_2$ (20 < x < 30) heterostructures. We replaced the commonly used AFM pinned layer with the ferrimagnet pinning layer. The interlayer thickness and FIM composition of the Tb–Co layer were optimized to obtain large H_{ex} and σ_{Pw}. The advantage of using amorphous RE-TM alloys as a pinned layer is the tunable magnetic properties that depend strongly on composition. The PEE reached a maximum σ_{Pw} = 1 (erg/cm^2) around x = 24 at.%, at room temperature. In this system, we observed two competing mechanisms—one leading to an increase and the other to a decrease—which corresponds to the effect of Tb content on saturation magnetization and the coercivity of heterostructures. The developed FIM/FM films, with a perpendicular exchange bias and a large PEE, will be greatly beneficial in spintronic applications, such as magneto-optical memory and high areal density recording technology.

Author Contributions: Conceptualization, S.R. and H.A.; investigation, S.R.; writing—original draft preparation, S.R.; writing—review and editing, S.R., S.S., K.T. and H.A. All authors have read and agreed to the published version of the manuscript.

Funding: This research received no external funding. This work was supported by the JSPS KAKENHI (grants 21K14202, and 20H02185).

Institutional Review Board Statement: Not applicable.

Informed Consent Statement: Not applicable.

Data Availability Statement: The data presented in this study are available upon reasonable request from the corresponding author.

Acknowledgments: The authors thank Ahmet Yagmur for his valuable discussions.

Conflicts of Interest: The authors declare no conflict of interest.

References

1. Meiklejohn, W.H.; Bean, C.P. New Magnetic Anisotorpy. *Phys. Rev.* **1957**, *105*, 904. [CrossRef]
2. Schuller, I.K. Exchange bias. *J. Magn. Magn. Mater.* **1999**, *192*, 203–232.
3. March, N.H.; Lambin, P.; Herman, F. Cooperative magnetic properties in single-and two-phase 3d metallic alloys relevant to exchange and magnetocrystalline anisotropy. *J. Magn. Magn. Mater.* **1984**, *44*, 1–19. [CrossRef]
4. Ranjbar, S.; Al-Mahdawi, M.; Oogane, M.; Ando, Y. High-Temperature Magnetic Tunnel Junction Magnetometers Based on L1 0 -PtMn Pinned Layer. *IEEE Sens. Lett.* **2020**, *4*, 5–8. [CrossRef]
5. Fujiwara, K.; Oogane, M.; Yokota, S.; Nishikawa, T.; Naganuma, H.; Ando, Y. Fabrication of magnetic tunnel junctions with a bottom synthetic antiferro-coupled free layers for high sensitive magnetic field sensor devices. *J. Appl. Phys.* **2012**, *111*, 5–8. [CrossRef]
6. Baibich, M.N.; Broto, J.M.; Fert, A.; Van Dau, F.N.; Petroff, F.; Eitenne, P.; Creuzet, G.; Friederich, A.; Chazelas, J. Giant magnetoresistance of (001)Fe/(001)Cr magnetic superlattices. *Phys. Rev. Lett.* **1988**, *61*, 2472–2475. [CrossRef]
7. Ranjbar, R.; Suzuki, K.; Sugihara, A.; Ma, Q.L.; Zhang, X.M.; Miyazaki, T.; Ando, Y.; Mizukami, S. Antiferromagnetic coupling in perpendicularly magnetized cubic and tetragonal Heusler bilayers. *Mater. Lett.* **2015**, *160*, 88–91. [CrossRef]
8. Yuasa, S.; Nagahama, T.; Fukushima, A.; Suzuki, Y.; Ando, K. Giant room-temperature magnetoresistance in single-crystal Fe/MgO/Fe magnetic tunnel junctions. *Nat. Mater.* **2004**, *3*, 868–871. [CrossRef]
9. Almeida, J.M.; Ferreira, R.; Freitas, P.P.; Langer, J.; Ocker, B.; Maass, W. 1f noise in linearized low resistance MgO magnetic tunnel junctions. *J. Appl. Phys.* **2006**, *99*, 08B314. [CrossRef]
10. Choi, S.; Lee, K.J.; Choi, S.; Chongthanaphisut, P.; Bac, S.K.; Lee, S.; Liu, X.; Dobrowolska, M.; Furdyna, J.K. Controllable Exchange Bias Effect in (Ga, Mn) As/(Ga, Mn)(As, P) Bilayers with Non-Collinear Magnetic Anisotropy. *IEEE Trans. Magn.* **2021**, *57*, 2–5. [CrossRef]
11. Suzuki, I.; Hamasaki, Y.; Itoh, M.; Taniyama, T. Controllable exchange bias in Fe/metamagnetic FeRh bilayers. *Appl. Phys. Lett.* **2014**, *105*, 172401. [CrossRef]
12. Sbiaa, R.; Piramanayagam, S.N. Multi-level domain wall memory in constricted magnetic nanowires. *Appl. Phys. A Mater. Sci. Process.* **2014**, *114*, 1347–1351. [CrossRef]
13. Atkinson, D.; Eastwood, D.S.; Bogart, L.K. Controlling domain wall pinning in planar nanowires by selecting domain wall type and its application in a memory concept. *Appl. Phys. Lett.* **2008**, *92*, 22510. [CrossRef]
14. Jung, H.S.; Traistaru, O.; Fujiwara, H. Effect of the kinds of ferromagnetic layers on exchange coupling strength in IrMn / FM films Effect of the kinds of ferromagnetic layers on exchange coupling strength in IrMn Õ FM films. *J. Appl. Phys.* **2004**, *95*, 6849–6851. [CrossRef]
15. Schmid, I.; Marioni, M.A.; Kappenberger, P.; Romer, S.; Parlinska-Wojtan, M.; Hug, H.J.; Hellwig, O.; Carey, M.J.; Fullerton, E.E. Exchange bias and domain evolution at 10 nm scales. *Phys. Rev. Lett.* **2010**, *105*, 197201. [CrossRef]
16. Tsunoda, M.; Nishikawa, K.; Damm, T.; Hashimoto, T.; Takahashi, M. Extra large unidirectional anisotropy constant of Co-Fe/Mn-Ir bilayers with ultra-thin antiferromagnetic layer. *J. Magn. Magn. Mater.* **2002**, *239*, 182–184. [CrossRef]
17. Ranjbar, S.; Tsunoda, M.; Oogane, M.; Ando, Y. Composition Dependence of Exchange Anisotropy in PtxMn1-x/Co70Fe30 Films. *Jpn. J. Appl. Phys.* **2019**, *58*, 043001. [CrossRef]
18. Cao, Y.; Rushforth, A.W.; Sheng, Y.; Zheng, H.; Wang, K. Tuning a Binary Ferromagnet into a Multistate Synapse with Spin–Orbit-Torque-Induced Plasticity. *Adv. Funct. Mater.* **2019**, *29*, 1808104. [CrossRef]
19. Ranjbar, S.; Tsunoda, M.; Al-mahdawi, M.; Oogane, M.; Ando, Y. Compositional Dependence of Exchange Anisotropy in PtxMn100−x/CoyFe100−y Films. *IEEE Magn. Lett.* **2019**, *10*, 1–5. [CrossRef]
20. Mangin, S.; Montaigne, F.; Schuhl, A. Interface domain wall and exchange bias phenomena in ferrimagnetic/ferrimagnetic bilayers. *Phys. Rev. B Condens. Matter Mater. Phys.* **2003**, *68*, 140404. [CrossRef]
21. Hebler, B.; Böttger, S.; Nissen, D.; Abrudan, R.; Radu, F.; Albrecht, M. Influence of the Fe-Co ratio on the exchange coupling in TbFeCo/[Co/Pt] heterostructures. *Phys. Rev. B* **2016**, *93*, 184423. [CrossRef]
22. Tokunaga, T.; Taguchi, M.; Fukami, T.; Nakaki, Y.; Tsutsumi, K. Study of interface wall energy in exchange-coupled double-layer film. *J. Appl. Phys.* **1990**, *67*, 4417–4419. [CrossRef]
23. Lin, C.C.; Lai, C.H.; Jiang, R.F.; Shieh, H.P.D. High interfacial exchange energy in TbFeCo exchange-bias films. *J. Appl. Phys.* **2003**, *93*, 6832–6834. [CrossRef]
24. Romer, S.; Marioni, M.A.; Thorwarth, K.; Joshi, N.R.; Corticelli, C.E.; Hug, H.J.; Oezer, S.; Parlinska-Wojtan, M.; Rohrmann, H. Temperature dependence of large exchange-bias in TbFe-Co/Pt. *Appl. Phys. Lett* **2012**, *101*, 222404. [CrossRef]
25. Radu, F.; Abrudan, R.; Radu, I.; Schmitz, D.; Zabel, H. Perpendicular exchange bias in ferrimagnetic spin valves. *Nat. Commun.* **2012**, *3*, 715. [CrossRef] [PubMed]
26. Hebler, B.; Reinhardt, P.; Katona, G.L.; Hellwig, O.; Albrecht, M. Double exchange bias in ferrimagnetic heterostructures. *Phys. Rev. B* **2017**, *95*, 104410. [CrossRef]
27. Canet, F.; Mangin, S.; Bellouard, C.; Piecuch, M. Positive exchange bias in ferromagnetic-ferrimagnetic bilayers: FeSn/FeGd. *Europhys. Lett.* **2000**, *52*, 594–600. [CrossRef]
28. Finley, J.; Liu, L. Spin-Orbit-Torque Efficiency in Compensated Ferrimagnetic Cobalt-Terbium Alloys. *Phys. Rev. Appl.* **2016**, *6*, 054001. [CrossRef]

29. Gottwald, M.; Hehn, M.; Montaigne, F.; Lacour, D.; Lengaigne, G.; Suire, S.; Mangin, S. Magnetoresistive effects in perpendicularly magnetized Tb-Co alloy based thin films and spin valves. *J. Appl. Phys.* **2012**, *111*, 083904. [CrossRef]
30. Siddiqui, S.A.; Han, J.; Finley, J.T.; Ross, C.A.; Liu, L. Current-Induced Domain Wall Motion in a Compensated Ferrimagnet. *Phys. Rev. Lett.* **2018**, *121*, 57701. [CrossRef]
31. Schubert, C.; Hebler, B.; Schletter, H.; Liebig, A.; Daniel, M.; Abrudan, R.; Radu, F.; Albrecht, M. Interfacial exchange coupling in Fe-Tb/[Co/Pt] heterostructures. *Phys. Rev. B Condens. Matter Mater. Phys.* **2013**, *87*, 054415. [CrossRef]
32. Shimanuki, S.; Ichihara, K.; Yasuda, N.; Ito, K.; Kohn, K. Magnetic and Magneto-Optical Properties of Amorphous TbCo Films Prepared by Two Target Magnetron Co-sputtering. *J. Magn. Soc. Jpn.* **1986**, *10*, 179–182. [CrossRef]
33. Tang, M.; Zhang, Z.; Jin, Q. Manipulation of perpendicular exchange bias effect in [Co/Ni]N/(Cu, Ta)/TbCo multilayer structures. *AIP Adv.* **2015**, *5*, 087153. [CrossRef]
34. Joo, S.J.; Hong, D.H.; Lee, T.D. Effect of Cu inserted layer between the IrMn and CoFeB interface on magnetic properties of CoFeB. *J. Appl. Phys.* **2004**, *95*, 7522–7524. [CrossRef]
35. Sheng, Y.; Edmonds, K.W.; Ma, X.; Zheng, H.; Wang, K. Adjustable Current-Induced Magnetization Switching Utilizing Interlayer Exchange Coupling. *Adv. Electron. Mater.* **2018**, *4*, 1800224. [CrossRef]
36. Tang, M.; Zhao, B.; Zhu, W.; Zhu, Z.; Jin, Q.Y.; Zhang, Z. Controllable Interfacial Coupling Effects on the Magnetic Dynamic Properties of Perpendicular [Co/Ni] 5 /Cu/TbCo Composite Thin Films. *ACS Appl. Mater. Interfaces* **2018**, *10*, 5090–5098. [CrossRef]
37. Tang, M.H.; Zhang, Z.; Tian, S.Y.; Wang, J.; Ma, B.; Jin, Q.Y. Interfacial exchange coupling and magnetization reversal in perpendicular [Co/Ni]N/TbCo composite structures. *Sci. Rep.* **2015**, *5*, 10863. [CrossRef]
38. Hansen, P.; Klahn, S.; Clausen, C.; Much, G.; Witter, K. Magnetic and magneto-optical properties of rare-earth transition-metal alloys containing Dy, Ho, Fe, Co. *J. Appl. Phys.* **1991**, *69*, 3194–3207. [CrossRef]
39. Harres, A.; Geshev, J. A polycrystalline model for magnetic exchange bias. *J. Phys. Condens. Matter* **2012**, *24*, 326004. [CrossRef]
40. Alebrand, S.; Gottwald, M.; Hehn, M.; Steil, D.; Cinchetti, M.; Lacour, D.; Fullerton, E.E.; Aeschlimann, M.; Mangin, S. Light-induced magnetization reversal of high-anisotropy TbCo alloy films. *Appl. Phys. Lett.* **2012**, *101*, 162408. [CrossRef]
41. Baruth, A.; Keavney, D.J.; Burton, J.D.; Janicka, K.; Tsymbal, E.Y.; Yuan, L.; Liou, S.H.; Adenwalla, S. Origin of the interlayer exchange coupling in [Co/Pt]/NiO/[Co/Pt] multilayers studied with XAS, XMCD, and micromagnetic modeling. *Phys. Rev. B* **2006**, *74*, 054419. [CrossRef]
42. Lederman, D.; Nogués, J.; Schuller, I.K. Exchange anisotropy and the antiferromagnetic surface order parameter. *Phys. Rev. B Condens. Matter Mater. Phys.* **1997**, *56*, 2332–2335. [CrossRef]
43. Kumar, D.; Singh, S.; Gupta, A. Effect of interface roughness on exchange coupling in polycrystalline Co/CoO bilayer structure: An in-situ investigation. *J. Appl. Phys.* **2016**, *120*, 085307. [CrossRef]
44. Hauet, T.; Mangin, S.; McCord, J.; Montaigne, F.; Fullerton, E.E. Exchange-bias training effect in TbFe GdFe: Micromagnetic mechanism. *Phys. Rev. B Condens. Matter Mater. Phys.* **2007**, *76*, 144423. [CrossRef]
45. Tsunoda, M.; Yoshitaki, S.; Ashizawa, Y.; Kim, D.Y.; Mitsumata, C.; Takahashi, M. Enhancement of exchange bias by ultra-thin Mn layer insertion at the interface of Mn-Ir/Co-Fe bilayers. *Phys. Status Solidi Basic Res.* **2007**, *244*, 4470–4473. [CrossRef]

magnetochemistry

![MDPI]

Article

Manipulation of Time- and Frequency-Domain Dynamics by Magnon-Magnon Coupling in Synthetic Antiferromagnets

Xing Chen [1,2], Cuixiu Zheng [1], Sai Zhou [1], Yaowen Liu [1,*] and Zongzhi Zhang [2,*]

1   Shanghai Key Laboratory of Special Artificial Microstructure Materials and Technology, School of Physics Science and Engineering, Tongji University, Shanghai 200092, China; 19110720009@fudan.edu.cn (X.C.); zhengcuixiu@tongji.edu.cn (C.Z.); 2010542@tongji.edu.cn (S.Z.)
2   Key Laboratory of Micro and Nano Photonic Structures (MOE), School of Information Science and Technology, Fudan University, Shanghai 200433, China
*   Correspondence: yaowen@tongji.edu.cn (Y.L.); zzzhang@fudan.edu.cn (Z.Z.)

Abstract: Magnons (the quanta of spin waves) could be used to encode information in beyond Moore computing applications. In this study, the magnon coupling between acoustic mode and optic mode in synthetic antiferromagnets (SAFs) is investigated by micromagnetic simulations. For a symmetrical SAF system, the time-evolution magnetizations of the two ferromagnetic layers oscillate in-phase at the acoustic mode and out-of-phase at the optic mode, showing an obvious crossing point in their antiferromagnetic resonance spectra. However, the symmetry breaking in an asymmetrical SAF system by the thickness difference, can induce an anti-crossing gap between the two frequency branches of resonance modes and thereby a strong magnon-magnon coupling appears between the resonance modes. The magnon coupling induced a hybridized resonance mode and its phase difference varies with the coupling strength. The maximum coupling occurs at the bias magnetic field at which the two ferromagnetic layers oscillate with a 90° phase difference. Besides, we show how the resonance modes in SAFs change from the in-phase state to the out-of-phase state by slightly tuning the magnon-magnon coupling strength. Our work provides a clear physical picture for the understanding of magnon-magnon coupling in a SAF system and may provide an opportunity to handle the magnon interaction in synthetic antiferromagnetic spintronics.

Keywords: magnons; synthetic antiferromagnets; antiferromagnetic resonance; micromagnetics

Citation: Chen, X.; Zheng, C.; Zhou, S.; Liu, Y.; Zhang, Z. Manipulation of Time- and Frequency-Domain Dynamics by Magnon-Magnon Coupling in Synthetic Antiferromagnets. *Magnetochemistry* 2022, 8, 7. https://doi.org/10.3390/magnetochemistry8010007

Academic Editor: Atsufumi Hirohata

Received: 1 December 2021
Accepted: 27 December 2021
Published: 30 December 2021

Publisher's Note: MDPI stays neutral with regard to jurisdictional claims in published maps and institutional affiliations.

1. Introduction

Magnon spintronics [1], which utilize propagating spin waves for nanoscale transmission and processing of information, have been growing as emerging research fields [2]. As a carrier of spin current, the magnons hold the promise of delivering information without the motion of electrons, therefore avoiding Ohmic losses and becoming a promising alternative to CMOS-based circuits. Recently, the magnons in antiferromagnets have attracted fundamental interest [3–5], in which long wavelength magnons can have frequency in the gigahertz (GHz), sub-terahertz, and even terahertz (THz) ranges because of the spin-sublattice exchange [6–8]. Due to the presence of two sublattices in antiferromagnets (AFMs) or ferrimagnets, the ferromagnetic resonance (FMR) spectra possess two different magnon branches. These two magnon modes can merge into a single branch at a degeneracy point, at which strong magnon-magnon interactions occur. This was reported in a layered AFM crystal CrCl$_3$ [9,10], compensated ferrimagnet gadolinium iron garnet [11], or magnetic metal-insulator hybrid structure [12,13]. Interestingly, the tunable magnon-magnon coupling in these systems provides the opportunity to use interactions between the magnon branches as a means to control/manipulate magnons in the device-based antiferromagnetic spintronics.

In contrast, synthetic antiferromagnets (SAFs) [14,15], composed of two ferromagnetic (FM) layers separated by a non-magnetic layer, could provide an easy way to handle the magnon-magnon interactions, because the interlayer coupling between the two FM layers mainly comes from the Ruderman–Kittel–Kasuya–Yosida (RKKY) interaction [16,17] and its strength is adjustable. The antiferromagnetically coupled two FM layers possess two kinds of uniform precession resonance modes: in-phase acoustic mode (AM) and out-of-phase optic mode (OM) [18–21]. Recently, it was shown that in a symmetrical SAF the symmetry-protected mode crossing in FMR spectra between the acoustic and optic mode branches can be eliminated by a tilting bias magnetic field [22,23] or by dynamic dipolar interaction from nonuniform precession of magnetic moments [24]. Subsequently, a strong magnon-magnon coupling between the AM and OM appears in the SAF systems. Besides, the SAFs with two asymmetric ferromagnetic sublayers (different thickness or different materials) can also break the system symmetry and realize the strong magnon-magnon coupling [23,25].

One of the important conditions to realize magnon-magnon coupling in the SAFs is to turn the two distant modes (AM and OM) into resonance at very close frequency, from which a new spin wave state can be generated, namely, mode hybridization. An obvious feature of the magnon hybridization state is the appearance of an anti-crossing gap in frequency spectra between the acoustic and optic branches. The strength of the magnon-magnon coupling can be characterized by the gap size [9]. Recently, the maximum coupling strength of 9.94 GHz was predicted [23]. However, most previous studies so far have focused on how to achieve magnon-magnon coupling or how to enhance the coupling strength, and less study is concerned with the magnetization precession of the hybrid modes. In this work, we will analytically and numerically study the magnetization dynamics of magnon-magnon coupling in both symmetrical and asymmetrical CoFeB-based SAFs. We find that a clear frequency crossing between the optic and acoustic magnon modes appear in a symmetrical SAF, indicating the absence of magnon coupling due to symmetry protection. However, for an asymmetrical SAF with different thicknesses of the two FM sublayers, a coupling gap is achieved because of the intrinsic symmetry breaking of the system. Remarkably, the strongest magnon-magnon coupling between the two magnon modes generates a hybrid precession mode with the phase difference of $\delta\varphi = 90°$ between the two magnetic sublattices, rather than the in-phase AM magnons ($\delta\varphi = 0°$) and out-of-phase OM magnons ($\delta\varphi = 180°$).

2. Simulation Model

As illustrated in Figure 1, we consider an SAF nanopillar structure of CoFeB (d_1 nm)/ Ru/CoFeB (d_2 nm) trilayer patterned in a circular shape of 100 nm × 100 nm. Here we consider two samples: One is symmetric SAF structure with $d_1 = d_2 = 2.0$ nm (Sample-I) and the other is an asymmetric SAF with $d_1 = 2.0$ nm and $d_2 = 4.0$ nm (Sample-II). In this study, the dynamics of the trilayer samples were simulated by using the open-source simulation software OOMMF (National Institute of Standards and Technology, Gaithersburg, MD, USA) [26], which is based on the Landau–Lifshitz–Gilbert equation:

$$\frac{d\boldsymbol{m_i}}{dt} = -\gamma\boldsymbol{m_i} \times \boldsymbol{H_{i,\text{eff}}} + \alpha\boldsymbol{m_i} \times \frac{d\boldsymbol{m_i}}{dt} \tag{1}$$

where $\boldsymbol{m_i} = \boldsymbol{M_i}/M_s$ is the unit magnetization vector of the ith discretization cell in upper or lower CoFeB layers. M_s is the saturation magnetization of CoFeB, α is the Gilbert damping factor, and γ is the gyromagnetic ratio. $\boldsymbol{H_{\text{eff}}}$ is the effective magnetic field that includes the intralayer exchange field, demagnetizing field $\boldsymbol{H_d}$, interlayer exchange field $\boldsymbol{H_{\text{IEC}}}$ between the upper and lower CoFeB, and external magnetic field $\boldsymbol{H_0}$. The effective magnetic field is:

$$\boldsymbol{H_{\text{eff}}} = \frac{2A_{ex}}{\mu_0 M_s}\nabla^2 \boldsymbol{m} + \boldsymbol{H_d} + \boldsymbol{H_{\text{IEC}}} + \boldsymbol{H_0} \tag{2}$$

where A_{ex} is the exchange stiffness, μ_0 is the vacuum permeability. $\boldsymbol{H_{\text{IEC}}} = J_{\text{IEC}}/(d_j M_s)$, here J_{IEC} is the interlayer exchange coupling constant, with $J_{\text{IEC}} > 0$ for ferromagnetic coupling while $J_{\text{IEC}} < 0$ for antiferromagnetic coupling. d_j is the thickness of CoFeB layer

(j = 1 or 2 refers to the upper or lower layer). In this study, we suppose the thickness of non-magnetic layer (d_{Ru}) is 1.1 nm (the second peak of antiferromagnetic coupling) [21]. Additionally, the typical material parameters of CoFeB (in CGS units) are used [27,28]: M_s = 1000 emu/cm^3, A_{ex} = 2.0 × 10^{-6} erg/cm, α = 0.01 and J_{IEC} = −0.2 erg/cm^2. Here, the magnetic anisotropy is ignored because its energy is almost unaffected by the in-plane orientation of the sublayer magnetizations [23,24]. All the simulations are performed without taking temperature into account.

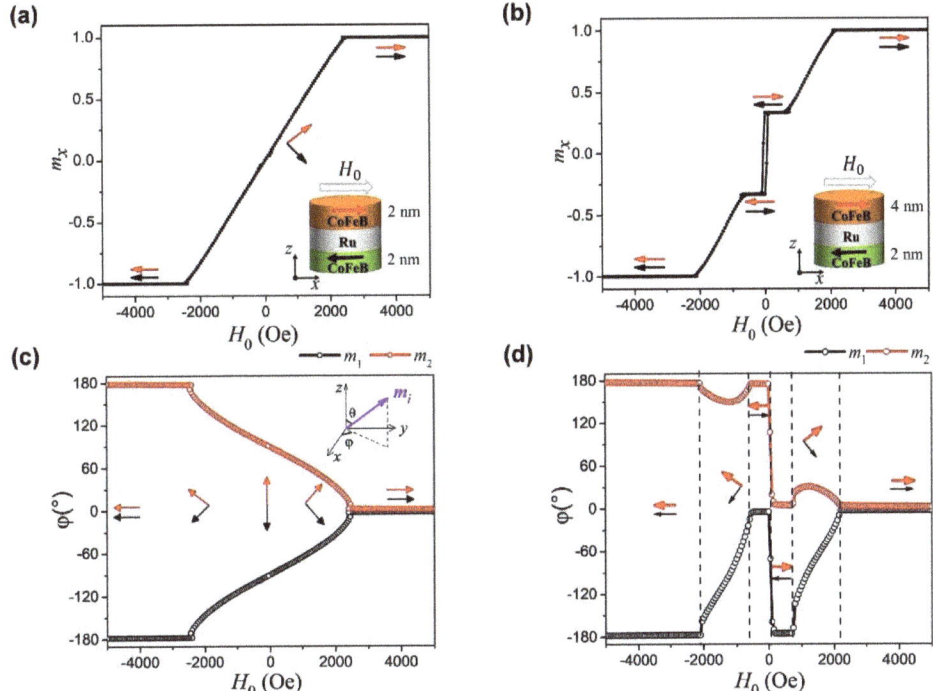

Figure 1. In-plane hysteresis loop as a function of the external field H_0 along x-direction for a symmetrical SAF (Sample-I) (**a**) and asymmetrical SAF (Sample-II) (**b**), respectively. The corresponding equilibrium angles of the magnetization vectors of two FM layers for the symmetrical SAF (**c**) and asymmetrical SAF (**d**) are also shown.

3. Results and Discussion

Firstly, we start our simulations for calculating the static hysteresis loops of two different samples and the results are shown in Figure 1. The external magnetic field H_0 is along the x-direction. For the symmetrical SAF (Sample-I), as shown in Figure 1a,c, there are only two equilibrium states: spin-canted state and parallel saturation state. Without the external magnetic field, the initial state of m_1 and m_2 are antiparallelly aligned due to the antiferromagnetic coupling. Under the action of H_0 with the strength of $0 < H_0 < H_s$, the m_1 and m_2 are rotated into a spin-canted state within the film plane, where $H_s = 2H_{IEC}$ represents the saturation magnetic field. The angles between m_i and H_0 satisfy: $\varphi_2 = -\varphi_1 = cos^{-1}(H_0/H_s)$. When $H_0 > H_s$, the m_1 and m_2 orient parallel to the direction of bias magnetic field (i.e., x-axis). Taking the parameter values of M_s, d, and J_{IEC}, we get the strength of H_{IEC} = 1100 Oe, which is in good agreement with the simulation results (the saturation field H_s = 2300 Oe and thereby $H_{IEC} = H_s/2$ = 1150 Oe).

For the asymmetrical SAF (Sample-II), however, as shown in Figure 1b,d, three typical equilibrium states exist. When the magnetic field is smaller than the critical field [18]

$H_{\text{cri},1} = H_{\text{IEC1}} - H_{\text{IEC2}}$, the magnetization vectors of m_1 and m_2 are opposite to each other along the x-direction and the net magnetization is constant. Above the critical field, the magnetizations of two FMs deviate from the antiparallel alignment. The field dependent magnetization can be estimated as [20,29]:

$$\frac{M(H)}{M_s} = \frac{d_1 \cos(\varphi_{1eq}) + d_2 \cos(\varphi_{2eq})}{d_1 + d_2} \tag{3}$$

where φ_{1eq} and φ_{2eq} represent the equilibrium directions of m_1 and m_2. When the external field H_0 is larger than $H_{\text{cri},2} = H_{\text{IEC1}} + H_{\text{IEC2}}$, both m_1 and m_2 are forced to align in the x-direction and the SAF reaches a saturation state.

3.1. Dynamic Resonance Properties of Symmetrical SAF

In this part, we study the dynamics of antiferromagnetic resonance. In addition to an external bias field applied along the x-axis to stabilize the magnetizations of SAFs, we also apply a time-varying microwave field of $h_{\text{rf}} = h_0 \sin(2\pi f t)$ to excite the magnons. The oscillation amplitude of the microwave fields is small and set as $h_0 = 3$ Oe in this study. Its direction is either along the x-axis (i.e., $h_{\text{rf}} \parallel H_0$, namely, longitudinal pumping) or along the y-axis (i.e., $h_{\text{rf}} \perp H_0$, namely, transverse pumping). The time evolution of spatially averaged magnetizations (m_1 and m_2) are recorded to calculate the response frequency through fast Fourier transform (FFT) analysis.

It is well known that an SAF system inherently has two distinct eigenmodes: AM and OM [18,19]. Figure 2 summarizes the simulated results at $H_0 = 600$ Oe for a symmetrical SAF with two identical FM layers ($d_1 = d_2 = 2.0$ nm). In the case of transverse pumping, only AM resonance is excited at a low frequency of 4 GHz while for the longitudinal pumping case only OM resonance is excited at a high frequency of 15 GHz, as shown in Figure 2a,b, respectively. This result can be well explained as follows: by considering that the resonance response signal is characterized by the rf components of the net magnetization $m = (m_1 + m_2)/2$ along the pumping field direction. For the low frequency resonance state, as shown in Figure 2c, d, both y- and z-components of m_1 and m_2 precess in phase while the x-component oscillates out-of-phase. In contrast, for the high frequency resonance state, as shown in Figure 2e,f, the x-component of m_1 and m_2 precess in-phase while both y- and z-components precess out-of-phase. Therefore, for the AFMR measurement with a transverse pumping microwave field (along the y-axis) to the bias magnetic field (x-axis), the in-phase AM resonance state (taken from the m_y or m_z) is only observed while the OM resonance state is hidden. In contrast, for the longitudinal pumping microwave field (along the x-axis), the observed resonance signal comes from the net magnetization m_x (because the net $m_y = 0$ or $m_z = 0$) but we classify this resonance state as the OM state due to the out-of-phase in m_y (or m_z) component.

Figure 3a shows the dispersion relation of frequency versus external magnetic field H_0 applied in the x-direction. In the spin-canted region, the frequency of the in-phase AM f_{AM} increases with the increasing field while the out-of-phase OM frequency f_{OM} decreases gradually until it reaches zero at the critical filed $H_s = 2300$ Oe. Theoretically, we could assume the whole FM layer is a single-domain and possesses a uniform magnetization precession within each layer. Thus, within the macrospin approximation, Equation (1) can be expanded as:

$$\begin{aligned}
\frac{dm_1}{dt} &= -\gamma m_1 \times \left(H_0 \vec{x} - H_{\text{IEC},1} m_2 - 4\pi M_s (m_1 \cdot \vec{z})\vec{z} \right) + \alpha m_1 \times \frac{dm_1}{dt} \\
\frac{dm_2}{dt} &= -\gamma m_2 \times \left(H_0 \vec{x} - H_{\text{IEC},2} m_1 - 4\pi M_s (m_2 \cdot \vec{z})\vec{z} \right) + \alpha m_2 \times \frac{dm_2}{dt}
\end{aligned} \tag{4}$$

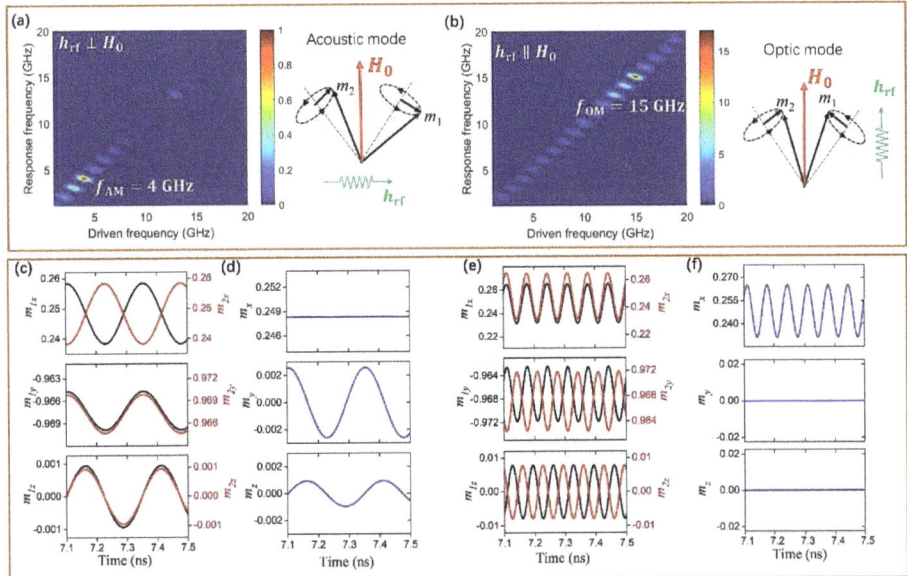

Figure 2. Top panel: The simulated AFMR response signal for the transverse pumping (**a**) and longitudinal pumping (**b**) at $H_0 = 600$ Oe. An acoustic resonance mode (AM) occurs at low frequency $f = 4$ GHz while the optic resonance mode (OM) occurs at $f = 15$ GHz. The color-coding refers to the resonance amplitude of magnetization response. The insert diagram shows the corresponding diagram of magnetization precession, where $m_1(m_2)$ represents the magnetization unit vector for ferromagnetic layer 1(2). Bottom panel: The magnetization oscillations of m_1 (black) and m_2 (red) in the acoustic mode (**c**) and optic mode (**e**), where m_y and m_z precess in phase in the acoustic mode while out of phase in the optic mode. The corresponding net magnetization $m = (m_1 + m_2)/2$ oscillations are also shown in (**d**) for AM and (**f**) for OM.

For any given applied field, Equation (4) has two real solutions, corresponding to the acoustic and optic modes. Additionally, for the symmetrical SAF ($d_1 = d_2 = 2.0$ nm), the angular frequencies of the two modes are expressed as [22]:

$$f_{AM} = \frac{\gamma}{2\pi} \sqrt{2H_{IEC}(2H_{IEC} + 4\pi M_s)} \frac{H_0}{2H_{IEC}}$$
$$f_{OM} = \frac{\gamma}{2\pi} \sqrt{8\pi H_{IEC} M_s \left(1 - \frac{H_0^2}{4H_{IEC}^2}\right)}$$

$$(5)$$

The theoretical results calculated from Equation (5) are plotted as the dotted black curve and red curve in Figure 3a, respectively. Clearly, the analytical result agrees well with the simulations.

A remarkable feature of the frequency spectra is the symmetry-protected mode crossing between the AM and OM branches, indicating that these two magnon modes have not hybridized in this symmetrical SAF. Consequently, no magnon-magnon coupling occurs. To further confirm the above results, the time-dependent phase difference between m_1 and m_2 near the crossing point are shown in Figure 3b,c, where φ_i ($i = 1, 2$) is the azimuth angle of magnetization vector. We find that m_1 and m_2 undergo pure in-phase precession in the AM magnon mode while anti-phase precession in the OM magnon mode. The simulation results are in agreement with the theoretical prediction.

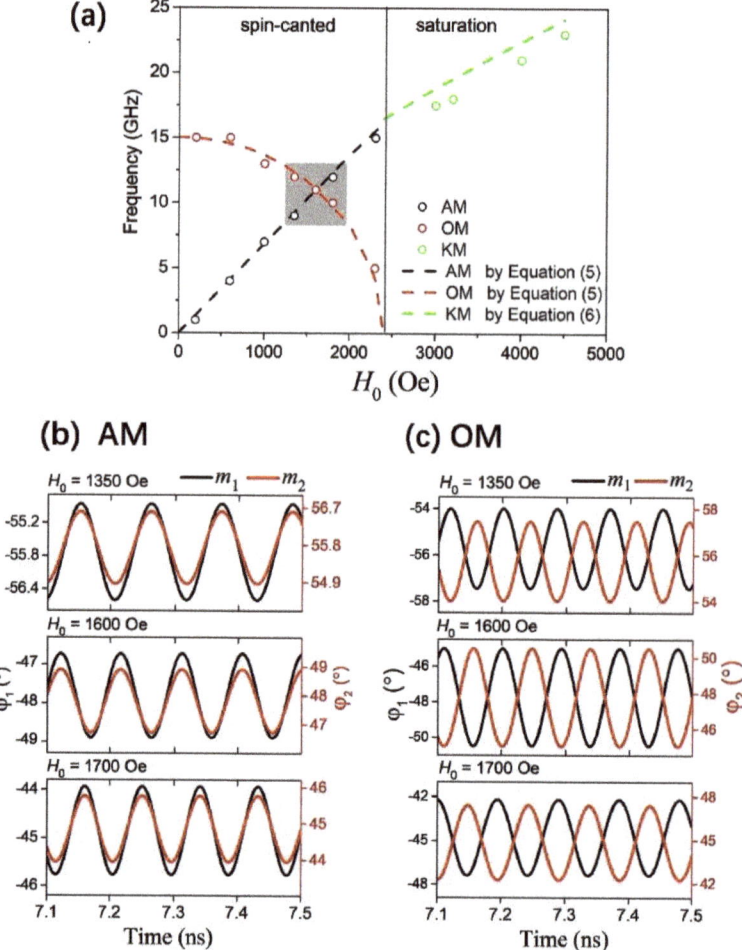

Figure 3. (a) Dispersion relation of frequency versus external magnetic field H_0 for the symmetrical SAF (Sample-I). The open circles represent the simulation results and dash lines represent the analytical calculations; (b,c) Comparison of the phase difference for the AM and OM resonant states at three different fields. Here φ_i ($i = 1, 2$) is the azimuth angle of m_i. It clearly shows the time-dependent in-phase precessions for the AM and antiphase precessions for the OM resonance state.

When the external field increases into the saturation region, both m_1 and m_2 are forced in the x-direction by the strong magnetic field. In this case, the SAF system behaves as a single ferromagnetic layer, the observed resonance state is the Kittel mode (KM) and its frequency can be described as [30]:

$$f_{\text{FMR}} = \frac{\gamma}{2\pi}\sqrt{H_0(H_0 + 4\pi M_s)} \tag{6}$$

In addition, the optic mode in this region will be hidden as its resonance intensity approaches to zero [31,32].

3.2. Dynamic Resonance Properties of Asymmetrical SAF

So far, we have investigated the magnon modes in symmetrical SAF with $d_1 = d_2$, and the magnon-magnon coupling does not occur due to symmetry-protection [9]. It has been theoretically predicted that the symmetry breaking will lead to a magnon-magnon coupling between the pure AM and OM, accompanied by an anti-crossing gap opening in frequency spectra [33]. This can be done in SAF structures by changing the two FM layers either with different materials or different thicknesses. To verify whether the intrinsic asymmetry can induce the coupling between the AM and OM, we simulated an asymmetrical SAF with different thicknesses, $d_1 = 2$ nm for the bottom layer and $d_2 = 4$ nm for the top layer. All other conditions are the same as the symmetrical SAF. The external bias field is applied along the x-direction.

Figure 4a shows the simulated dispersion relation of the asymmetrical SAF. An obvious anti-crossing gap is observed in the spin-canted region, indicating the magnon-magnon coupling phenomena appears. Here we define the magnon coupling strength as $g = (f_{\text{up}} - f_{\text{down}})/2$, where f_{up} and f_{down} refer to the minimum frequency of up branch and the maximum frequency of down branch. The simulation shows that the magnon-magnon coupling strength is $g = 1.5$ GHz in our sample. Theoretically, the resonance frequency can also be derived from the eigenvalue equation of Equation (4) [33]:

$$\omega^4 - \left(a^2 + \frac{c}{2b}\right)\omega^2 + \frac{2b+c}{4b^2}(1-a^2)(a^2+c-1) = 0 \qquad (7)$$

where $a = \frac{H_0}{H_{cri,2}}$, $b = \frac{2|J_{IEC}|}{\mu_0 M_s^2 (d_1+d_2)}$, and $c = 1 - \left(\frac{d_1-d_2}{d_1+d_2}\right)^2$. Thus, the angular frequency of the up and down branches can be obtained as:

$$\omega_{\text{up,down}} = \frac{1}{\sqrt{2}}\sqrt{\left(a^2 + \frac{c}{2b}\right) \pm \frac{\sqrt{b^2+2b+c}}{b}\sqrt{\left[a^2 - \frac{(2b+c)(2-c)-bc}{2(b^2+2b+c)}\right]^2 + \frac{(1-c)(2b+c)^3}{4(b^2+2b+c)^2}}} \qquad (8)$$

The theoretical curves calculated from Equation (8) are depicted by dash lines in Figure 4a in the spin-canted region.

To acquire a clearer insight into the behavior of magnon-magnon coupling, we further studied the phase difference of magnetization evolution between m_1 and m_2, as shown in Figure 4b,c. The down mode and up mode have obvious changes with the external magnetic field. As the external magnetic field H_0 increases, the down mode changes from pure AM to OM, while the up mode changes from pure OM to AM. Remarkably, the phase difference of the new hybrid mode is not 0° or 180° but almost 90° at the strongest coupling field ($H_0 = 1500$ Oe). Actually, this process of change can also establish the relationship between the phase difference and the magnon coupling strength.

So far, we have investigated the magnon–magnon coupling in symmetrical SAF with $d_1 = d_2$ and asymmetrical SAF with $d_1 \neq d_2$. For symmetrical SAF, only pure in-phase AM and out-of-phase OM are observed and the phase difference $\delta\varphi = \varphi_1 - \varphi_2$ between m_1 and m_2 is constant ($\delta\varphi = 0°$ for the AM and 180° for the OM magnons), as shown in Figure 5a. For the asymmetrical SAF, as shown in Figure 5b, the phase difference $\delta\varphi$ varies with the external magnetic field H_0, showing the $\delta\varphi$ changes from 0° to 180° for the down magnon mode while changes from 180° to 0° for the up mode. An obvious hybrid characteristic is shown.

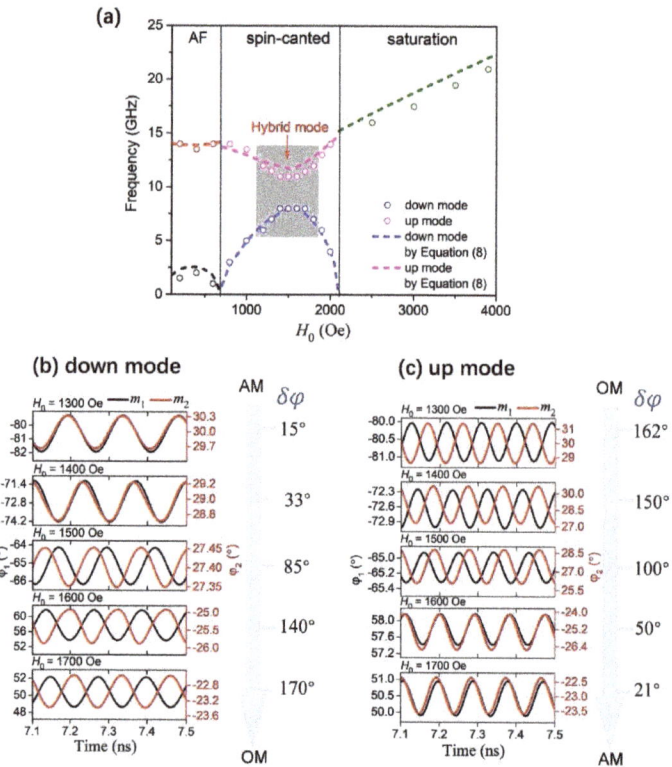

Figure 4. (**a**) Dispersion relation of frequency in asymmetrical SAF (Sample-II) with the external magnetic field H_0 applied along the *x*-axis direction. The open circles represent the simulation results while dotted lines represent the corresponding theoretical calculations, showing an obvious coupling gap. Phase difference of down mode (**b**) and up mode (**c**) at 1300 Oe, 1400 Oe,1500 Oe, 1600 Oe, and 1700 Oe. The phase difference between $\mathbf{m}_1$ and $\mathbf{m}_2$ varies with the external magnetic field, indicating a new hybrid mode.

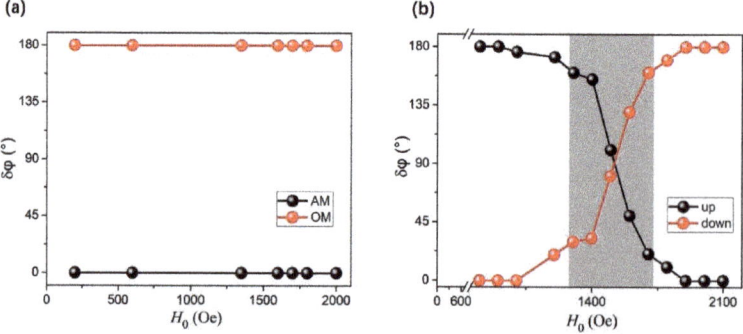

Figure 5. (**a**) Phase difference of pure AM (black curve) and pure OM (red curve) for the symmetrical SAF (Sample-I); (**b**) Phase difference of down branch mode (red curve) and up branch mode (black curve) for the asymmetrical SAF (Sample-II).

4. Conclusions

In summary, we have investigated the coupling between acoustic and optic magnon modes in both symmetrical and asymmetrical SAFs. In addition to the frequency dispersion, here we pay more attention to the phase difference between the sublayer magnetizations for these two SAF systems. For the symmetrical SAF, the in-phase AM and out-of-phase OM exist separately and their magnetic field-dependent frequency branches cross each other at a degenerate point, and no magnon-magnon coupling occurs. In contrast, for an asymmetrical SAF, however, an obvious anti-crossing gap appears in frequency dispersion relations and the magnon-magnon coupling between the AM and OM occurs due to the intrinsic symmetry breaking of the system. The original AM and OM magnons gradually hybridizes with the increase of the coupling strength. The phase difference between m_1 and m_2 is almost 90° at the strongest coupling field. The study demonstrates a clear physical picture of the mode coupling, from which a hybrid spin-wave mode can be generated by turning the two distinct modes into resonance and the physical properties of the coupled modes have changed significantly. In addition, generating the hybrid spin-wave states may also provide assistance in the development of future magnonic devices.

Author Contributions: Conceptualization, X.C. and Y.L.; data curation, C.Z. and S.Z.; formal analysis, X.C., Y.L. and Z.Z.; writing—original draft preparation, X.C.; writing—review and editing, Y.L. and Z.Z. All authors have read and agreed to the published version of the manuscript.

Funding: This work is supported by the National Key Research and Development Project of China (2018YFB0407603) and the National Natural Science Foundation of China (Grant Nos. 11774260, 51971161, 11874120, 52171230).

Institutional Review Board Statement: Not applicable.

Informed Consent Statement: Not applicable.

Data Availability Statement: The study did not report any data.

Conflicts of Interest: The authors declare no conflict of interest. The funders had no role in the design of the study; in the collection, analyses, or interpretation of data; in the writing of the manuscript, or in the decision to publish the results.

References

1. Chumak, A.V.; Vasyuchka, V.I.; Serga, A.A.; Hillebrands, B. Magnon spintronics. *Nat. Phys.* **2015**, *11*, 453. [CrossRef]
2. Yu, H.; Xiao, J.; Schultheiss, H. Magnetic texture based magnonics. *Phys. Rep.* **2021**, *905*, 1. [CrossRef]
3. Cheng, R.; Xiao, J.; Niu, Q.; Brataas, A. Spin Pumping and Spin-Transfer Torques in Antiferromagnets. *Phys. Rev. Lett.* **2014**, *113*, 057601. [CrossRef] [PubMed]
4. Baltz, V.; Manchon, A.; Tsoi, M.; Moriyama, T.; Ono, T.; Tserkovnyak, Y. Antiferromagnetic spintronics. *Rev. Mod. Phys.* **2018**, *90*, 015005. [CrossRef]
5. Rezende, S.M.; Azevedo, A.; Rodríguez-Suárez, R.L. Introduction to antiferromagnetic magnons. *J. Appl. Phys.* **2019**, *126*, 151101. [CrossRef]
6. Cheng, R.; Xiao, D.; Brataas, A. Terahertz Antiferromagnetic Spin Hall Nano-Oscillator. *Phys. Rev. Lett.* **2016**, *116*, 207603. [CrossRef]
7. Vaidya, P.; Morley, S.A.; van Tol, J.; Liu, Y.; Cheng, R.; Brataas, A.; Lederman, D.; del Barco, E. Subterahertz spin pumping from an insulating antiferromagnet. *Science* **2020**, *368*, 160. [CrossRef]
8. Li, J.; Wilson, C.B.; Cheng, R.; Lohmann, M.; Kavand, M.; Yuan, W.; Aldosary, M.; Agladze, N.; Wei, P.; Sherwin, M.S.; et al. Spin current from sub-terahertz-generated antiferromagnetic magnons. *Nature* **2020**, *578*, 70. [CrossRef]
9. MacNeill, D.; Hou, J.T.; Klein, D.R.; Zhang, P.; Jarillo-Herrero, P.; Liu, L. Gigahertz Frequency Antiferromagnetic Resonance and Strong Magnon-Magnon Coupling in the Layered Crystal $CrCl_3$. *Phys. Rev. Lett.* **2019**, *123*, 047204. [CrossRef]
10. Sklenar, J.; Zhang, W. Self-Hybridization and Tunable Magnon-Magnon Coupling in van der Waals Synthetic Magnets. *Phys. Rev. Appl.* **2021**, *15*, 044008. [CrossRef]
11. Liensberger, L.; Kamra, A.; Maier-Flaig, H.; Geprägs, S.; Erb, A.; Goennenwein, S.T.B.; Gross, R.; Belzig, W.; Huebl, H.; Weiler, M. Exchange-Enhanced Ultrastrong Magnon-Magnon Coupling in a Compensated Ferrimagnet. *Phys. Rev. Lett.* **2019**, *123*, 117204. [CrossRef]
12. Li, Y.; Cao, W.; Amin, V.P.; Zhang, Z.; Gibbons, J.; Sklenar, J.; Pearson, J.; Haney, P.M.; Stiles, M.D.; Bailey, W.E.; et al. Coherent Spin Pumping in a Strongly Coupled Magnon-Magnon Hybrid System. *Phys. Rev. Lett.* **2020**, *124*, 117202. [CrossRef]

13. Chen, J.; Liu, C.; Liu, T.; Xiao, Y.; Xia, K.; Bauer, G.E.W.; Wu, M.; Yu, H. Strong Interlayer Magnon-Magnon Coupling in Magnetic Metal-Insulator Hybrid Nanostructures. *Phys. Rev. Lett.* **2018**, *120*, 217202. [CrossRef]
14. Parkin, S.S.P.; More, N.; Roche, K.P. Oscillations in exchange coupling and magnetoresistance in metallic superlattice structures: Co/Ru, Co/Cr, and Fe/Cr. *Phys. Rev. Lett.* **1990**, *64*, 2304. [CrossRef]
15. Grunberg, P.; Schreiber, R.; Pang, Y.; Brodsky, M.B.; Sowers, H. Layered magnetic structures: Evidence for antiferromagnetic coupling of Fe layers across Cr interlayers. *Phys. Rev. Lett.* **1986**, *57*, 2442. [CrossRef]
16. Ruderman, M.A.; Kittel, C. Indirect Exchange Coupling of Nuclear Magnetic Moments by Conduction Electrons. *Phys. Rev.* **1954**, *96*, 99. [CrossRef]
17. Kasuya, T. A Theory of Metallic Ferro- and Antiferromagnetism on Zener's Model. Prog. *Theor. Phys.* **1956**, *16*, 45. [CrossRef]
18. Zhang, Z.; Zhou, L.; Wigen, P.E.; Ounadjela, K. Angular dependence of ferromagnetic resonance in exchange-coupled Co/Ru/Co trilayer structures. *Phys. Rev. B* **1994**, *50*, 6094. [CrossRef] [PubMed]
19. Rezende, S.M.; Chesman, C.; Lucena, M.A.; Azevedo, A.; de Aguiar, F.M.; Parkin, S.S.P. Studies of coupled metallic magnetic thin-film trilayers. *J. Appl. Phys.* **1998**, *84*, 958. [CrossRef]
20. Belmeguenai, M.; Martin, T.; Woltersdorf, G.; Maier, M.; Bayreuther, G. Frequency- and time-domain investigation of the dynamic properties of interlayer-exchange-coupled $Ni_{81}Fe_{19}/Ru/Ni_{81}Fe_{19}$ thin films. *Phys. Rev. B* **2007**, *76*, 104414. [CrossRef]
21. Waring, H.J.; Johansson, N.A.B.; Vera-Marun, I.J.; Thomson, T. Zero-field Optic Mode Beyond 20 GHz in a Synthetic Antiferromagnet. *Phys. Rev. Appl.* **2020**, *13*, 034035. [CrossRef]
22. Sud, A.; Zollitsch, C.W.; Kamimaki, A.; Dion, T.; Khan, S.; Iihama, S.; Mizukami, S.; Kurebayashi, H. Tunable magnon-magnon coupling in synthetic antiferromagnets. *Phys. Rev. B* **2020**, *102*, 100403. [CrossRef]
23. Dai, C.; Ma, F. Strong magnon–magnon coupling in synthetic antiferromagnets. *Appl. Phys. Lett.* **2021**, *118*, 112405. [CrossRef]
24. Shiota, Y.; Taniguchi, T.; Ishibashi, M.; Moriyama, T.; Ono, T. Tunable Magnon-Magnon Coupling Mediated by Dynamic Dipolar Interaction in Synthetic Antiferromagnets. *Phys. Rev. Lett.* **2020**, *125*, 017203. [CrossRef] [PubMed]
25. He, W.; Xie, Z.K.; Sun, R.; Yang, M.; Li, Y.; Zhao, X.-T.; Liu, W.; Zhang, Z.D.; Cai, J.-W.; Cheng, Z.-H.; et al. Anisotropic Magnon–Magnon Coupling in Synthetic Antiferromagnets. *Chin. Phys. Lett.* **2021**, *38*, 057502. [CrossRef]
26. Donahue, M.J.; Porter, D.G. *OOMMF User's Guide*; Interagency Report NISTIR 6376; NIST: Gaithersburg, MD, USA, 1999. Available online: http://math.nist.gov/oommf (accessed on 10 October 2021).
27. Kanai, S.; Yamanouchi, M.; Ikeda, S.; Nakatani, Y.; Matsukura, F.; Ohno, H. Electric field-induced magnetization reversal in a perpendicular-anisotropy CoFeB-MgO magnetic tunnel junction. *Appl. Phys. Lett.* **2012**, *101*, 122403. [CrossRef]
28. Devolder, T.; Bianchini, L.; Miura, K.; Ito, K.; Kim, J.-V.; Crozat, P.; Morin, V.; Helmer, A.; Chappert, C.; Ikeda, S.; et al. Spin-torque switching window, thermal stability, and material parameters of MgO tunnel junctions. *Appl. Phys. Lett.* **2011**, *98*, 162502. [CrossRef]
29. Sorokin, S.; Gallardo, R.A.; Fowley, C.; Lenz, K.; Titova, A.; Atcheson, G.Y.P.; Dennehy, G.; Rode, K.; Fassbender, J.; Lindner, J.; et al. Magnetization dynamics in synthetic antiferromagnets: Role of dynamical energy and mutual spin pumping. *Phys. Rev. B* **2020**, *101*, 144410. [CrossRef]
30. Kittel, C. On the Theory of Ferromagnetic Resonance Absorption. *Phys. Rev.* **1948**, *73*, 155. [CrossRef]
31. Chen, X.; Zheng, C.; Zhang, Y.; Zhou, S.; Liu, Y.; Zhang, Z. Identification and manipulation of spin wave polarizations in perpendicularly magnetized synthetic antiferromagnets. *New J. Phys.* **2021**, *23*, 113029. [CrossRef]
32. Chen, X.; Zheng, C.; Zhou, S.; Liu, Y.; Zhang, Z. Ferromagnetic resonance modes of a synthetic antiferromagnet at low magnetic fields. *J. Phys. Condens. Matt.* **2021**, *34*, 015802. [CrossRef]
33. Li, M.; Lu, J.; He, W. Symmetry breaking induced magnon-magnon coupling in synthetic antiferromagnets. *Phys. Rev. B* **2021**, *103*, 064429. [CrossRef]

magnetochemistry

Article

Theory of Antiferromagnet-Based Detector of Terahertz Frequency Signals

Ansar Safin [1,2,*], **Sergey Nikitov** [1,3], **Andrei Kirilyuk** [1,4], **Vasyl Tyberkevych** [5] and **Andrei Slavin** [5]

1   Kotel'nikov Institute of Radioengineering and Electronics, Russian Academy of Sciences, 125009 Moscow, Russia; nikitov@cplire.ru (S.N.); andrei.kirilyuk@ru.nl (A.K.)
2   Department of Radioengineering and Electronics, Moscow Power Engineering Institute, National Research University, 111250 Moscow, Russia
3   Moscow Institute of Physics and Technology, Dolgoprudny, 141700 Moscow, Russia
4   FELIX Laboratory, Radboud University, 6525 AJ Nijmegen, The Netherlands
5   Department of Physics, Oakland University, Rochester, MI 48309, USA; tyberkev@oakland.edu (V.T.); slavin@oakland.edu (A.S.)
*   Correspondence: arsafin@gmail.com

Citation: Safin, A.; Nikitov, S.; Kirilyuk, A.; Tybekevych, V.; Slavin, A. Theory of Antiferromagnet-Based Detector of Terahertz Frequency Signals. *Magnetochemistry* **2022**, *8*, 26. https://doi.org/10.3390/magnetochemistry8020026

Academic Editor: Raymond F. Bishop

Received: 31 December 2021
Accepted: 8 February 2022
Published: 12 February 2022

Publisher's Note: MDPI stays neutral with regard to jurisdictional claims in published maps and institutional affiliations.

Abstract: We present a theory of a detector of terahertz-frequency signals based on an antiferromagnetic (AFM) crystal. The conversion of a THz-frequency electromagnetic signal into the DC voltage is realized using the inverse spin Hall effect in an antiferromagnet/heavy metal bilayer. An additional bias DC magnetic field can be used to tune the antiferromagnetic resonance frequency. We show that if a *uniaxial* AFM is used, the detection of linearly polarized signals is possible only for a non-zero DC magnetic field, while circularly polarized signals can be detected in a zero DC magnetic field. In contrast, a detector based on a *biaxial* AFM can be used without a bias DC magnetic field for the rectification of both linearly and circularly polarized signals. The sensitivity of a proposed AFM detector can be increased by increasing the magnitude of the bias magnetic field, or by by decreasing the thickness of the AFM layer. We believe that the presented results will be useful for the practical development of tunable, sensitive and portable spintronic detectors of THz-frequency signals based of the antiferromagnetic resonance (AFMR).

Keywords: spin pumping; spin-orbit torque; insulating antiferromagnet; sub-terahertz waves; spin-Hall effect

1. Introduction

Frequency-selective and tunable detection of terahertz (THz) frequency signals is an operation that is important for many different applications—from medical scanning, to security, to high-speed 6G communication and radio astronomy [1]. Due to the rarity of resonators with natural frequencies in the THz (from 0.1 to 10 THz) frequency range, the tunable resonance detection in this frequency range is still a significant challenge [2–5]. One option to realize resonance detection of THz-frequency signals is to use antiferromagnetic (AFM) crystals that naturally have frequencies of the antiferromagnetic resonance (AFMR) in the THz-frequency range. These high frequencies of the AFMR are related to the existence of a strong exchange interaction between the AFM magnetic sublattices (internal exchange magnetic fields of up to 10^2–10^3 T) [6].

It has been shown theoretically that AFMs can be used as active layers of THz-frequency oscillators [7–10] and detectors [11–13]. Recent experiments on the effect of spin-pumping performed in both uniaxial [14–16] and biaxial [17,18] AFMs indicate the possibility of development of THz frequency-detectors based on antiferromagnet/heavy metal (AFM/HM) heterostructures. In this work, we analyze the available theoretical and experimental data on the properties of AFM crystals, and describe the influence of the AFM crystal anisotropy, magnitude and orientation of the external bias magnetic field, as well as the polarization of the received THz-frequency electromagnetic signal, on the possibility of resonance detection of such signals using spin pumping in passive spintronic detectors-based AFM/HM bilayers.

The general theory of spin-pumping and spin-transfer torque in AFM/HM layered structures was developed in [19]. The influence of the signal polarization and the type of the AFM anisotropy on the detection of THz-frequency signals by the AFM/HM spintronic detectors has been further studied theoretically in [11,13]. It was found that a *uniaxial* AFM gives a zero rectified voltage for a *linearly* polarized AC spin current signal, but can detect a *circularly* polarized AC signal [13]. It was also found that a *biaxial* dielectric AFM (such as NiO) can be used as a sensitive element of a resonance quadratic rectifier of *linearly* polarized AC spin current signals, and that a sensitivity of such a rectifier could be in the range of 1 kV/W [11,13]. The conditions necessary for using *uniaxial* AFMs for the detection of *linearly* polarized signals have not been studied in detail, so far.

It is well-known (see, e.g., [6]) that, in the absence of an external bias magnetic field, the AFMR frequencies in AFM crystals are proportional to the square root of the product of the anisotropy fields and the AFM internal exchange field. The AFM internal exchange magnetic field, which keeps the AFM sublattices anti-parallel to each other, reaches hundreds of Tesla, while the AFM anisotropy field is much smaller (from μT to several T), and, therefore, the tuning of the AFMR frequency is, usually, done by changing the AFM anisotropy fields. The variation of the anisotropy fields can be done using magnetostriction in the adjacent piezoelectric layer [20,21], driving DC current through the adjacent HM layer [13], or by changing temperature [22]. When an external bias magnetic field is applied to a uniaxial AFM, its influence on the AFMR frequency depends on the field direction relative to the anisotropy easy axis, and linear tuning of the AFMR frequency is possible when the bias field is parallel to the anisotropy easy axis, but the bias field magnitude necessary for the AFMR tuning is rather large, of the order of several tesla.

In this work, we consider a theory of resonance detection of both *linearly* and *circularly* polarized electromagnetic (EM) signals via a spin-pumping mechanism in AFM/HM heterostructures. We assume the presence of a DC external bias magnetic field that can be used for tuning the AFMR frequency of the detector, as it was done in recent AFM spin-pumping experiments [14–16]. We also study the additional influence of the bias magnetic field on the detector properties. The paper is organized as follows. In Section 2, we describe the possible physical structure of an AFM/HM based detector. In Section 3, we present a mathematical model of the magnetization dynamics in an AFM using the so-called "sigma-model" developed in [23–25] for both uniaxial and biaxial AFM crystals. The expressions for the AFM-based detector sensitivity are presented in Section 4, while the conclusions are given in Section 5.

2. Physical Structure

Let us consider a concept of a THz-frequency detector based on AFM/HM bilayer, which is shown in Figure 1a. Here, the anisotropy easy-axis is oriented in the plane of the sample $\mathbf{e}_{EA} = \mathbf{e}_3$. The magnetic field component of the AC electromagnetic field $\mathbf{h}_{AC} = h_{AC}\mathbf{e}_{AC} \cdot e^{i\omega t}$ created by an external signal is oriented in the plane perpendicular to the easy-axis $\mathbf{e}_{EA}$, where $\mathbf{e}_{AC} = (\mathbf{e}_1 + \mathbf{e}_2)/\sqrt{2}$ and $\mathbf{e}_{AC} = (\mathbf{e}_1 + i\mathbf{e}_2)/\sqrt{2}$ for the cases of linear (LP) and circular (CP) polarization, respectively, while h_{AC} and ω are the amplitude and frequency of the AC magnetic field.

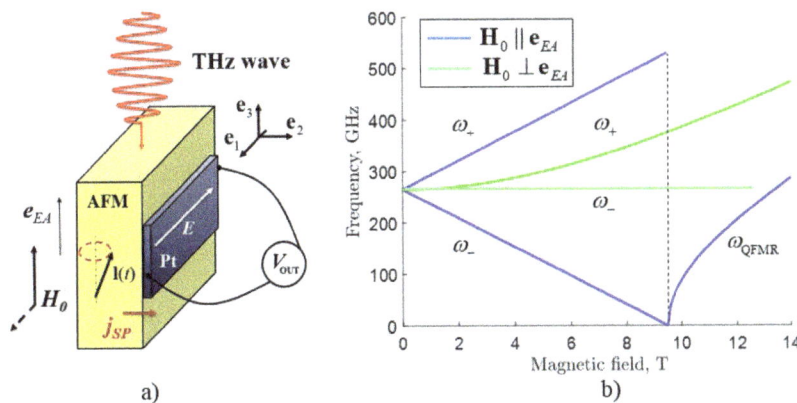

a)   b)

Figure 1. (**a**) Schematic view of the resonance detector based on an AFM/HM heterostructure under the action of a THz-frequency electromagnetic signal with controllable polarization. Here l(t) is the Neel vector oriented along the anisotropy easy-axis, V_{OUT} is the output DC electric voltage, and $\mathbf{H_0}$ is the external DC bias magnetic field; (**b**) dependence of the resonance frequencies of the detector based on the uniaxial AFM MnF_2 on the DC bias magnetic field H_0 for $\mathbf{H_0} \parallel \mathbf{e}_{EA}$ and $\mathbf{H_0} \perp \mathbf{e}_{EA}$.

The external AC magnetic field induces torque, which acts on the magnetic sublattices of the AFM, and causes oscillations of the Néel vector $\mathbf{l} = (\mathbf{M_1} - \mathbf{M_2})/2M_s$ near the easy-axis, and creates a spin-current due to the spin-pumping mechanism [6]:

$$\mathbf{j}_{SP} = \frac{\hbar g_r}{2\pi}\left[\mathbf{l} \times \frac{d\mathbf{l}}{dt}\right], \tag{1}$$

where g_r is the real part of the spin-mixing conductance, $\hbar$ is the reduced Planck constant, $\mathbf{M_{1,2}}$ are magnetization vectors of the AFM sublattices, and M_s is the saturation magnetization of the sublattices. This spin current is then injected into the HM, which produces a charge current and electric field between the output electrodes E through the inverse spin Hall effect (ISHE), and results in an electric DC voltage V_{OUT}. The experimental ISHE voltages and other physical parameters of different uniaxial and biaxial AFMs for the zero external DC magnetic field are presented in Table 1, and they are all above tens of nV. One can see from Table 1 the resonance frequencies of the presented AFMs lie in the THz frequency range. We use MnF_2 and NiO for our numerical simulations for uniaxial and biaxial cases, respectively, as materials with low damping at room temperatures, which give the acceptable quality factor for AFM resonance.

Table 1. Parameters of uniaxial and biaxial AFMs at zero DC magnetic field $H_0 = 0$.

Material	$H_{EA,HA}$, T	H_{ex}, T	f_{AFMR}, GHz	Δf, GHz	V_{ISHE}, nV	Ref.
FeF_2	20	108	1400	30	-	[26,27]
Cr_2O_3	0.07	490	163	5.6	30	[15,17]
MnF_2	0.85	106	245	2.6	60	[14,16]
NiO	0.03, 0.7668	1937	220, 1100	18	-	[28,29]

3. Magnetization Precession Induced by a Polarized THz EM Signal

A general phenomenological method for the description of the AFM dynamics is based on the use of coupled Landau–Lifshitz equations for the magnetizations of the sublattices $\mathbf{M_{1,2}}$ [6]. Using this approach under the condition that the total magnetization $\mathbf{M} = \mathbf{M_1} + \mathbf{M_2}$ is small, I.V. Baryakhtar and B.A. Ivanov [23] obtained an effectively closed equation describing the dynamics of an antiferromagnet in terms of a normalized (unit) vector $\mathbf{l} = (\mathbf{M}_1 - \mathbf{M}_2)/2M_s$. In their derivation, it was assumed that the magnetization vector M of an antiferromagnet is a "slave" variable, and is determined by the vector $\mathbf{l}(t)$ and its time derivative $d\mathbf{l}(t)/dt$. The dynamic equations of motion for the unit vector $\mathbf{l}(t)$ are usually called the equations of the "sigma-model", and their application greatly simplifies the analysis of both linear and nonlinear dynamic effects in antiferromagnets [6]. A.F. Andreev and V.I. Marchenko [25], as well as A.K. Zvezdin [24], obtained the sigma-model equation based on the analysis of the dynamic symmetry of the AFM. In this section, we describe the magnetization dynamics in an AFM crystal using the sigma-model in the following form [23,25]:

$$\mathbf{l} \times \left(\frac{d^2\mathbf{l}}{dt^2} + \gamma_{\text{eff}} \frac{d\mathbf{l}}{dt} - 2\gamma \left[\frac{d\mathbf{l}}{dt} \times \mathbf{H}_0 \right] + \frac{\partial W_{\text{AFM}}}{\partial \mathbf{l}} \right) = \left[\mathbf{l} \times \gamma \frac{d\mathbf{h}_{\text{AC}}}{dt} \right] \times \mathbf{l}, \qquad (2)$$

where $\gamma_{\text{eff}} = \alpha_{\text{eff}}\omega_{\text{ex}}$ is the spectral linewidth of the AFM resonance at zero bias magnetic field H_0 [13], α_{eff} is the effective damping including Gilbert constant and spin-pumping term [11], $\gamma = 2\pi \cdot 28$ GHz/T is the gyromagnetic ratio. The vector product $d\mathbf{l}/dt \times \mathbf{H}_0$ is the gyroscopic torque [25] and $W_{\text{AFM}}(\mathbf{l}, \mathbf{H}_0)$ is the magnetic energy density in the presence of the DC bias magnetic field, which can be expressed in the form (see for more details [23,25]):

$$W_{\text{AFM}}(\mathbf{l}, \mathbf{H}_0) = -\frac{\omega_{\text{ex}}\omega_{\text{EA}}}{2}(\mathbf{l} \cdot \mathbf{e}_{\text{EA}})^2 + \frac{\omega_{\text{ex}}\omega_{\text{HA}}}{2}(\mathbf{l} \cdot \mathbf{e}_{\text{HA}})^2 + \frac{\gamma^2}{2}(\mathbf{H}_0 \cdot \mathbf{l})^2. \qquad (3)$$

Here characteristic frequencies are defined as follows: $\omega_{\text{ex}} = \gamma H_{\text{ex}}$, $\omega_{\text{EA}} = \gamma H_{\text{EA}}$, $\omega_{\text{HA}} = \gamma H_{\text{HA}}$, and H_{ex} is the AFM internal exchange magnetic field, $H_{\text{EA}}, H_{\text{HA}}$ are the AFM anisotropy fields corresponding to the easy and hard axes, respectively (see Table 1). Some authors use a definition of the exchange field in an AFM, $H_E = H_{\text{ex}}/2$, which is half of the exchange field H_{ex} used in our current work. We use the definition $H_{\text{ex}} = 2 \cdot H_E$. following the classical papers on the magnetization dynamics in AFM crystals [23,25]. Thus, the left-hand side part in Equation (2) contains the inertial, damping, gyroscopic, and anisotropy terms, respectively, while the right-hand side part of the equation describes the influence of the AC magnetic field of the external signal. Note, that in [11,13] an AC spin current with a torque $[\mathbf{l} \times \mathbf{j}_{\text{AC}}] \times \mathbf{l}$ in the right-hand side of Equation (2) was used as an excitation mechanism, where $\mathbf{j}_{\text{AC}}$ is the density of the spin-current. Our further results on the study of model (2) with external electromagnetic radiation are also applicable to the case of a spin current.

Let us now consider the small-amplitude dynamics of the Néel vector expressed as a sum of the static component $\mathbf{l}_0$, describing the AFM ground state, and a small dynamic vector $\mathbf{s}(t)$ excited by the AC magnetic field of the external signal:

$$\mathbf{l}(t) = \mathbf{l}_0 + \mathbf{s}(t). \qquad (4)$$

Note, that the vectors $\mathbf{l}_0$ and $\mathbf{s}$ satisfy the orthogonality constraint, i.e., $(\mathbf{l}_0 \cdot \mathbf{s}) = 0$. The ansatz (4) uses the assumption of a small change in the dynamic vector $\mathbf{s}(t)$ near the stationary vector $\mathbf{l}_0$, which describes the AFM ground state. This is a common technique in the theory of oscillations and waves. In such a linear theory it is assumed that the vector $\mathbf{s}(t)$ is small, and we can linearize the original nonlinear sigma-model equation to obtain a linear dynamic equation for the vector $\mathbf{s}(t)$. In a nonlinear case, the ansatz $\mathbf{l} = \mathbf{l}_0 + \mathbf{s}$ can also be used, but, then, Equation (6) must be modified, and the nonlinear terms must be considered in that equation. Such nonlinear dynamics can include the second harmonic generation [30], or the appearance of the self-oscillations [8], but the theory of such nonlinear processes is beyond the scope of our current manuscript.

The equation defining the AFM ground state Néel vector $\mathbf{l}_0$ can be easily found from (3) as follows:

$$\omega_{\text{ex}}\omega_{\text{EA}}(\mathbf{l}_0 \cdot \mathbf{e}_{\text{EA}})[\mathbf{l}_0 \times \mathbf{e}_{\text{EA}}] - \omega_{\text{ex}}\omega_{\text{HA}}(\mathbf{l}_0 \cdot \mathbf{e}_{\text{HA}})[\mathbf{l}_0 \times \mathbf{e}_{\text{HA}}] = \omega_H^2(\mathbf{l}_0 \cdot \mathbf{e}_H)[\mathbf{l}_0 \times \mathbf{e}_H], \qquad (5)$$

where $\omega_H = \gamma H_0$, and e_H is the unit vector along the DC bias magnetic field. Solving Equation (5) gives the ground state Néel vector $l_0 = e_3$.

Using Equation (4) in Equation (2) we can derive the following differential equation describing the oscillations of the dynamic part of the Néel vector $s = s_1 e_1 + s_2 e_2$:

$$\frac{d^2 s}{dt^2} + \gamma_{eff}\frac{ds}{dt} - 2\omega_H(l_0 \cdot e_H) \cdot \hat{\Theta} \cdot \frac{ds}{dt} + \left(\hat{\Omega} - (l_0 \cdot \hat{\Omega}l_0)\hat{I}\right) \cdot s = \hat{\Theta} \cdot \gamma\frac{dh_{AC}}{dt}, \qquad (6)$$

where matrices $\hat{\Theta}, \hat{I}, \hat{\Omega}$ can be expressed as follows:

$$\hat{\Theta} = \begin{pmatrix} 0 & 1 \\ -1 & 0 \end{pmatrix}, \hat{I} = \begin{pmatrix} 1 & 0 \\ 0 & 1 \end{pmatrix}, \qquad (7)$$

$$\hat{\Omega} = -\omega_{ex}\omega_{EA}e_{EA} \otimes e_{EA} + \omega_{ex}\omega_{HA}e_{HA} \otimes e_{HA} + \omega_H^2 e_H \otimes e_H. \qquad (8)$$

Linear vectorial Equation (6) describes the small-amplitude dynamics of the AFM Néel vector. The formal solution $s(\omega)$ of Equation (6) for a harmonic driving signal $h_{AC} = \omega_{AC}e_{AC}e^{i\omega t}$ (here $\omega_{AC} = \gamma h_{AC}$) has the following form:

$$s(\omega) = i\omega\omega_{AC}\hat{D}^{-1}(\omega) \cdot \hat{\Theta} \cdot e_{AC}, \qquad (9)$$

where $\hat{D}(\omega)$ is the matrix

$$\hat{D}(\omega) = \left[\left(-\omega^2 + i\gamma_{eff}\omega\right)\hat{I} - 2i\omega\omega_H(l_0 \cdot e_H)\hat{\Theta} + (\hat{\Omega} - (l_0 \cdot \hat{\Omega}l_0)\hat{I})\right]. \qquad (10)$$

We can rewrite expression (9) in the form:

$$\begin{pmatrix} s_1 \\ s_2 \end{pmatrix} = \frac{i\omega\omega_{AC}}{\det[\hat{D}(\omega)]}\begin{pmatrix} \omega_2^2 - \omega^2 + i\omega\gamma_{eff} & 2i\omega\omega_H(e_H \cdot l_0) \\ -2i\omega\omega_H(e_H \cdot l_0) & \omega_1^2 - \omega^2 + i\omega\gamma_{eff} \end{pmatrix} \cdot \begin{pmatrix} e_{AC,2} \\ -e_{AC,1} \end{pmatrix}, \qquad (11)$$

where $e_{AC,1,2} = 1/\sqrt{2}$ for LP and $e_{AC,1} = i/\sqrt{2}$, $e_{AC,1} = 1/\sqrt{2}$ for CP, and

$$\omega_{1,2}^2 = \omega_{ex}(\omega_{EA} + \omega_{HA}(e_{1,2} \cdot e_{HA})) + \omega_H^2 \cdot \left((e_{1,2} \cdot e_H)^2 - (e_3 \cdot e_H)^2\right). \qquad (12)$$

Now, we can find a general expression for the AFM eigenfrequencies $\omega_\pm$ in the case of zero effective damping γ_{eff}. These eigenfrequencies are found from the condition of the vanishing of the determinant of the matrix (10) in the following form (here we take $\overline{\omega}_H = \omega_H : (e_H \cdot l_0)$):

$$\omega_\pm^2 = \frac{1}{2}\left(\omega_1^2 + \omega_2^2\right) + 2 \cdot \overline{\omega}_H^2 \pm \sqrt{\frac{1}{4}\left(\omega_1^2 - \omega_2^2\right)^2 + 2 \cdot \overline{\omega}_H^2(\omega_1^2 + \omega_2^2) + 4\overline{\omega}_H^4} \qquad (13)$$

Let us consider several particular cases for the orientation of the external bias magnetic field H_0 relative to the axes $e_{1,2,3}$ in the uniaxial and biaxial AFM crystals.

(a) *Easy-axis uniaxial AFM* ($H_{HA} = 0$).

For the case of a zero DC bias magnetic field, two eigenfrequencies $\omega_\pm$ are degenerate, and equal to $\omega_\pm = \sqrt{\omega_{ex}\omega_{EA}} = \omega_{AFMR}^0$. The dynamic vector $s(\omega)$ has, in this case, the simplest form:

$$\begin{pmatrix} s_1 \\ s_2 \end{pmatrix} = \frac{i\omega\omega_{AC}}{(-\omega^2 + (\omega_{AFMR}^0)^2 + i\omega\gamma_{eff})} \cdot \begin{pmatrix} e_{AC,2} \\ -e_{AC,1} \end{pmatrix}. \qquad (14)$$

This is a standard expression for the amplitude–frequency characteristic of an oscillatory system with one degree of freedom FM modes are (two AFM modes $s_{1,2}$ are degenerate and uncoupled).

In the case when $\mathbf{H}_0 \parallel \mathbf{e}_{EA}$, the resonance frequencies from (13) can be found in the following form:

$$\omega_\pm = \omega_{AFMR}^0 \pm \omega_H \quad \text{for} \quad H_0 < H_{sf},$$ (15)

and

$$\omega_{QFMR} = \sqrt{(\omega_H)^2 - (\omega_{AFMR}^0)^2} \quad \text{for} \quad H_0 > H_{sf},$$ (16)

where $H_{sf} = \sqrt{H_{ex} \cdot H_{EA}}$ is the spin-flop field, at which the Néel vector changes its direction from the parallel to the external bias magnetic field to the perpendicular to it. The dependences of the resonant frequencies defined by the expressions (15) and (16) are shown in Figure 1b. Such dependences were obtained experimentally for different easy-axis AFMs (see, e.g., [15,16]). Since the rectification of the modes having "quasi-ferromagnetic" frequency requires a bias field higher than the field of a spin-flop transition (which for MnF_2 is 9.4 T, and for Cr_2O_3 is 6 T), and, therefore, requires the use of sources of rather large magnetic fields, in the following we shall restrict our attention to the rectification of signals in bias fields below the spin-flop transition.

In the case when $\mathbf{H}_0 \perp \mathbf{e}_{EA}$, the AFMR frequencies are:

$$\omega_+ = \sqrt{\left(\omega_{AFMR}^0\right)^2 + \omega_H^2}, \quad \omega_- = \omega_{AFMR}^0,$$ (17)

The upper frequency quadratically increases with the increase of the DC bias magnetic field, while the lower mode frequency is constant and equal to ω_{AFMR}^0.

(b) Easy-plane biaxial AFM ($H_{HA} \neq 0$).

For the zero DC bias magnetic field two AFM frequencies $\omega_\pm$ are non-degenerate and equal to $\sqrt{\omega_{ex}\omega_{EA}}$ and $\sqrt{\omega_{ex}(\omega_{HA} + \omega_{EA})}$. Most often, the hard-axis field H_{HA} is much larger than the easy-axis field H_{EA} (see Table 1 for the nickel oxide), and the effect of the easy-plane anisotropy variation on the higher resonance frequency can be neglected. Qualitatively, the nature of the dependences shown in Figure 1b coincides for the easy-axis and the easy-plane cases.

In the particular case when $\mathbf{H}_0 \parallel \mathbf{e}_{EA}$ and $\mathbf{e}_{HA} = \mathbf{e}_1$ the resonance frequencies are equal to (before the spin-flop field [29]):

$$\omega_+ \approx \sqrt{\omega_{ex}\omega_{HA} + 3\omega_H^2}, \quad \omega_- \approx \sqrt{\omega_{ex}\omega_{EA} - \omega_H^2}.$$ (18)

For the $\mathbf{H}_0 \perp \mathbf{e}_{EA}$ one of the AFMR frequencies does not depend on the magnetic field, and the second one grows quadratically.

Let us now study the influence of the driving AC signal polarization on the rectified DC voltage in AFM obtained as a result of the spin pumping for various relative orientations between the direction of the external bias DC magnetic field and the anisotropy axes.

4. Rectification of THz-Frequency Electromagnetic Signals

Let us derive an expression for the inverse spin Hall DC voltage V_{OUT} induced by the spin pumping from the AFM into the adjacent HM layer. Using (1) and (4) we get this expression in the following form:

$$V_{OUT} = \kappa \langle s_1 \frac{ds_2}{dt} - s_2 \frac{ds_1}{dt} \rangle = 2i\omega\kappa[s_1^* s_2 - s_2^* s_1],$$ (19)

where κ is the proportionality coefficient

$$\kappa = \frac{L g_r \theta_{SH} e \lambda_{Pt} \rho}{2\pi d_{Pt}} \tanh\left(\frac{d_{Pt}}{2\lambda_{Pt}}\right),$$ (20)

L is the distance between output electrodes, θ_{SH} is the spin-Hall angle, e is the electron charge, λ_{Pt} is the spin-diffusion length, while ρ and d_{Pt} are the electrical resistivity and thickness of the Pt layer, respectively. For the input AC power of the EM signal $P_{AC} = \frac{c}{2\mu_0} S \cdot (h_{AC})^2$, where c is the speed of light, μ_0 is the magnetic permeability, S is the AFM layer cross-section, one can find the detector sensitivity defined as:

$$R(\omega) = \frac{|V_{OUT}(\omega)|}{P_{AC}} \qquad (21)$$

For the cases of linear and circular polarizations of the driving AC signal we get the detector sensitivities as:

$$R_{LP}(\omega) = R_0 \cdot \frac{\omega_{ex}\omega^3 |(\omega_1^2 - \omega^2)(\omega\gamma_{eff} + 2\omega\overline{\omega}_H) - (\omega_2^2 - \omega^2)(\omega\gamma_{eff} - 2\omega\overline{\omega}_H)|}{|\det(\hat{D}(\omega))|^2}, \qquad (22)$$

$$R_{CP}(\omega) = R_0 \cdot \frac{\omega_{ex}\omega^3 |(\omega_1^2 - \omega^2 - 2\omega\overline{\omega}_H)(\omega_2^2 - \omega^2 - 2\omega\overline{\omega}_H) + (\omega\gamma_{eff})^2|}{|\det(\hat{D}(\omega))|^2}, \qquad (23)$$

where $R_0 = 4\kappa\gamma^2\mu_0/(S\omega_{ex}c)$.

Now, let us analyze the above obtained expressions (22) and (23) for detector sensitivity in two different cases of uniaxial and biaxial AFM crystals.

(a) Easy-axis uniaxial AFM ($H_{HA} = 0$).

The rectified output DC voltage is equal to zero due to the fact, that the modes $s_{1,2}$ are uncoupled for the LP in both cases $H_0 = 0$ and $\mathbf{H}_0 \perp \mathbf{e}_{EA}$. When $\mathbf{H}_0 \parallel \mathbf{e}_{EA}$, two modes $s_{1,2}$ are mutually coupled due to the gyroscopic terms in Equation (2), and a non-zero sensitivity can be obtained from (22):

$$R_{LP}(\omega) = \frac{4R_0\omega_{ex}\omega^4 |\omega_1^2 - \omega^2|\omega_H}{|\det \hat{D}(\omega)|^2}. \qquad (24)$$

One can see from the expression (24) that the detector sensitivity is proportional to the bias DC magnetic field $H_0 = 0$. Figure 2a shows the resonance-type dependence of the sensitivity on the frequency ω for the upper branch ω_+ of the resonance curve shown in Figure 1a in the case of a non-zero external DC bias magnetic field. In our numerical calculations, we assumed that the AFM layer is made of MnF$_2$, and used the following coefficients taken from [16]: $\alpha_{eff} = 0.5 \times 10^{-3}$, $\theta_{SH} = 0.08$, $\lambda_{Pt} = 1.4$ nm, $d_{Pt} = 5$ nm, $\rho = 2.5 \times 10^{-7}$ $\Omega \cdot$ m, $g_r = 2.86 \times 10^{18}$ m^{-2}, $d_{AFM} = 10$ nm, $L = 100$ μm. As can be seen from Figure 2a and Equation (18), the resonance sensitivity increases with the increase of the the bias magnetic field H_0. Note, that the input AC power of the EM signal is defined as $P_{AC} = \frac{c}{2\mu_0} S \cdot (h_{AC})^2$, so for the AC signal amplitude $h_{AC} = 0.1$ mT and the AFM cross-section $S = 100 \times 100$ nm^2 we get the value of $P_{AC} = 12$ nW. The dependence of the detector sensitivity on the bias magnetic field for linearly polarized (LP) and circularly polarized (CP) signals is shown in Figure 2b for the above given parameters and nano-scale sizes of the AFM/HM heterostructure. When the magnitude of the DC magnetic field is varied, the resonance frequency shifts, as it is shown in Figure 1b, while the spectral linewidth of a resonance curve remains unchanged, as it is equal to $\alpha_{eff} \cdot \omega_{ex}$. In the recent experiment [15,16] performed in bulk mm-size AFM samples the observed detector sensitivity was near $10^{-5} - 10^{-6}$ V/W, which is quite small compared to our above presented theoretical estimation made for a nano-sized AFM sample. We believe that the main reason for this huge difference is the relatively large size of the AFM layer used in [15,16]. It has been also theoretically demonstrated recently [11,13] that the sensitivity of an AFM detector can reach several kV/W for detectors using nanometer-thick AFM layers. As follows from expressions (19) and (20), the output voltage of an AFM detector is inversely proportional to the AFM thickness. As it was shown in [11], when the AFM thickness decreases, there is an optimal AFM thickness at which the sensitivity reaches the maximum value. With a further decrease of the AFM thickness, the sensitivity decreases. To correctly calculate the sensitivity at thicknesses of the order of several nanometers, it is necessary to use the modified sigma model (2), in which the additional spatial derivatives are included. This calculation was presented in [11], and it is not repeated in our current work. Thus, we come to the obvious conclusion that the nano-sized sensitive AFM elements should be used in the future design of the spintronic AFM detectors of THz-frequency EM signals. Another possible way to increase the detector sensitivity is to use several nano-scale detectors mutually coupled through a common HM layer, or to use magnetic tunnel junctions to extract the output voltage [10].

For a circularly polarized EM signal in both cases $\mathbf{H}_0 \parallel \mathbf{e}_{EA}$ and $\mathbf{H}_0 \perp \mathbf{e}_{EA}$, one can get a non-zero diode sensitivity described by the equation:

$$R_{CP}(\omega) = \frac{R_0 \omega_{ex} \omega^3 |(\omega_1^2 - \omega^2 - 2\omega\overline{\omega}_H)^2 + (\omega\gamma_{eff})^2|}{|\det \hat{D}(\omega)|^2}. \tag{25}$$

The rectification of a THz signal at a zero bias magnetic field was studied earlier in [13], where the driving THz-frequency signal had the form of a spin-polarized current. However, the presence of an external bias DC magnetic field removes the degeneracy of the eigen-frequencies of the system, and increases the magnitude of the rectified voltage. Additionally, the use of a driving signal with circular polarization makes possible the observation of the rectified spin-pumping voltage both in the presence, and in the absence of an external bias magnetic field. In contrast, in the case of a linear polarization of the driving signal, such an observation is realized only for $\mathbf{H}_0 \parallel \mathbf{e}_A$. As can be seen from Figure 2, the sensitivity for CP signals is larger than for the LP signals at the same value of the DC magnetic field. In the CP case, the expression (25) consists of two terms: one is linearly proportional to the DC magnetic field, while the other one is independent of it. In contrast, in the LP case the sensitivity expression (24) contains only one term proportional to the DC magnetic field. The summary for the calculation of eigen-frequencies $\omega_{\pm}$ and sensitivity at different ratios between the orientation of the external magnetic field is presented in Table 2.

Table 2. Expressions for the eigen-frequencies $\omega_{\pm}$ and AFM diode sensitivity R at different orientations of the external bias magnetic field $\mathbf{H}_0$ relative to the easy axis $\mathbf{e}_{EA}$ of the AFM layer and polarizations (LP or CP) of the external EM signal for the uniaxial AFM crystal. The numbers in parentheses ($\cdot$) correspond to the equation numbers in the main text of the paper.

Parameter	$\mathbf{H}_0 = 0$	$\mathbf{H}_0 \parallel \mathbf{e}_{EA}$	$\mathbf{H}_0 \perp \mathbf{e}_{EA}$
$\omega_{\pm}$	$\sqrt{\omega_{ex}\omega_{EA}}$	$\sqrt{\omega_{ex}\omega_{EA}} \pm \omega_H$	$\sqrt{\omega_{ex}\omega_{EA}}, \sqrt{\omega_{ex}\omega_{EA} + \omega_H^2}$
R_{LP}	0	(24)	0
R_{CP}	(25)	(25)	(25)

(b) Easy-plane biaxial AFM ($H_{HA} \neq 0$).

It was shown previously [11] that a biaxial AFM can be used to rectify a *linearly* polarized AC spin current in the case when the AFM easy plane is oriented *perpendicular* to the plane of the AFM sample. The maximum value of the rectified voltage is achieved when the angle between the direction of the spin-current polarization and the directions of the AFM anisotropy axes is 45 degrees. In this case, it is possible to obtain a non-zero rectified voltage even in a *zero* bias DC magnetic field.

At the same time, from the technological point of view, it is easier to fabricate biaxial AFM crystal in the case when the easy plane coincides with the plane of the sample, or is inclined to the sample plane at a small angle. In this work, we consider only the situation when the AFM easy plane *coincides* with the plane of the AFM sample.

For the determination of sensitivity in the case of a biaxial AFM one needs to use the general expressions (22) and (23). The analysis presented above for the uniaxial non-degenerate case is applicable to the biaxial case as well. The resonance curve for the NiO is shown in Figure 2c, and is characterized by two resonance eigen-frequencies even in a zero DC magnetic field. The dependence of the diode sensitivity on the bias magnetic field obtained in such a case is shown in Figure 2d. The sensitivity of the lowest-frequency mode in a zero bias magnetic field and at a linear polarization of the external AC signal is much smaller than for the case of a circular polarization (0.1 V/W for LP and 27 V/W for CP), but it is, in general, non-zero. In our numerical calculations we used the physical parameters for the NiO crystal taken from [28]. It is easy to see that in the case of a biaxial AFM (similar to the above discussed case of a uniaxial AFM), the increase of the DC bias magnetic field leads to the increase of the diode sensitivity.

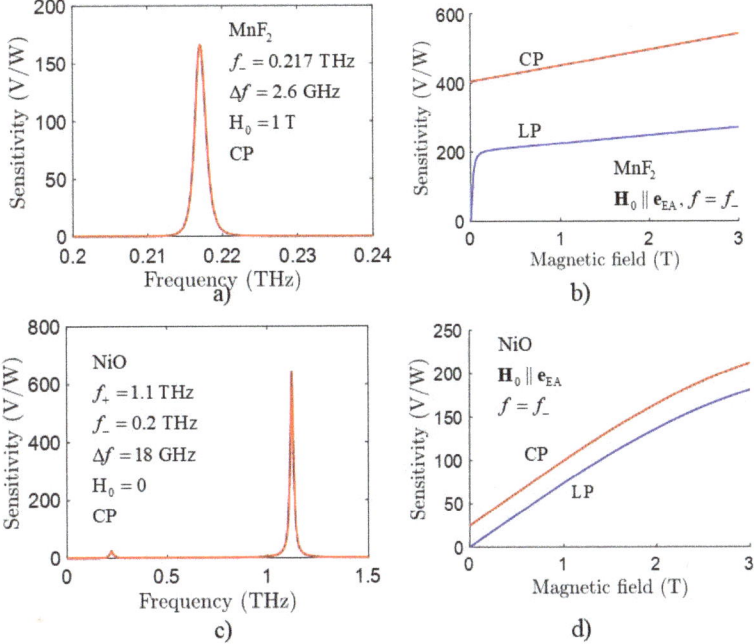

Figure 2. AFM detector sensitivity as a function of the external signal frequency (**a**,**c**) and bias magnetic field (**b**,**d**) for the AFM-HM heterstructures containing uniaxial (MnF_2) and biaxial (NiO) AFM crystals. Calculation were performed for the lowest-frequency AFMR modes).

5. Conclusions

We demonstrated theoretically that nanometer-size AFM/HM heterostructures can be used for the development of sensitive and tunable resonance detectors of THz frequency signals. We found that: (i) Using *uniaxial* AFM crystals the detection of *linearly* polarized electromagnetic signals is possible only for a non-zero DC bias magnetic field, while the signals having *circular* (or elliptical) polarization could be detected even in a zero bias magnetic field; (ii) using *biaxial* AFM crystals it is possible to detect both *linearly* and *circularly* polarized EM signals in a zero bias magnetic field, but in the presence of a bias magnetic field field the detector sensitivity increases with the increase of the bias field for both uniaxial and biaxial AFM crystals; (iii) to increase the sensitivity of an AFM detector it is necessary to decrease the thickness of the sensitive AFM element, since the detection mechanism is based on the interface spin-Hall effect. We believe that our results will be useful for the development of tunable and highly sensitive THz-frequency AFM devices controlled by an applied bias DC magnetic field, such as spectrum analyzers [31] or/and neuromorphic signal processors [28,32].

Author Contributions: Conceptualization, A.S. (Ansar Safin), S.N., A.K., V.T. and A.S. (Andrei Slavin); mathematical model, A.S. (Andrei Slavin) and V.T., data curation, A.S. (Andrei Slavin); formal analysis, A.S. (Ansar Safin), A.S. (Andrei Slavin) and V.T.; writing original draft preparation, A.S. (Ansar Safin); writing review and editing, A.S. (Ansar Safin), S.N., A.K., V.T. and A.S. (Andrei Slavin). All authors have read and agreed to the published version of the manuscript.

Funding: This work was partially funded by the Russian Science Foundation (Grant No. 21-79-10396), U.S. National Science Foundation (Grant # EFMA-1641989), the Air Force Office of Scientific Research under the MURI grant # FA9550-19-1-0307, from the DARPA TWEED grant # DARPA-PA-19-04-05-FP-001, and from the Oakland University Foundation.

Institutional Review Board Statement: Not applicable.

Informed Consent Statement: Not applicable.

Data Availability Statement: The study did not report any data.

Conflicts of Interest: The authors declare no conflict of interests. The funders had no role in the design of the study; in the collection, analysis, or interpretation of data; in the writing of the manuscript, or in the decision to publish the results.

References

1. Dhillon, S.S.; Vitiello, M.S.; Linfield, E.H.; Davies, A.G.; Hoffmann, M.C.; Booske, J.; Paoloni, C.; Gensch, M.; Weightman, P.; Williams, G.P.; et al. The 2017 terahertz science and technology roadmap. *J. Phys. Appl. Phys.* **2017**, *50*, 043001. [CrossRef]
2. Jiang, S.L.; Jia, X.Q.; Jin, B.B.; Kang, L.; Xu, W.W.; Chen, J.; Wu, P.H. Superconducting detectors for terahertz imaging. In Proceedings of the 2015 40th International Conference on Infrared, Millimeter, and Terahertz waves (IRMMW-THz), Hong Kong, China, 23–28 August 2015. [CrossRef]
3. Spirito, D.; Coquillat, D.; Bonis, S.L.D.; Lombardo, A.; Bruna, M.; Ferrari, A.C.; Pellegrini, V.; Tredicucci, A.; Knap, W.; Vitiello, M.S. High performance bilayer-graphene terahertz detectors. *Appl. Phys. Lett.* **2014**, *104*, 061111. [CrossRef]
4. Viti, L.; Cadore, A.R.; Yang, X.; Vorobiev, A.; Muench, J.E.; Watanabe, K.; Taniguchi, T.; Stake, J.; Ferrari, A.C.; Vitiello, M.S. Thermoelectric graphene photodetectors with sub-nanosecond response times at terahertz frequencies. *Nanophotonics* **2020**, *10*, 89–98. [CrossRef]
5. Sanchez-Martin, H.; Sanchez-Martin, S.; de-la Torre, I.I.; Perez, S.; Novoa, J.A.; Ducournau, G.; Grimbert, B.; Gaquiere, C.; Gonzalez, T.; Mateos, J. GaN nanodiode arrays with improved design for zero-bias sub-THz detection. *Semicond. Sci. Technol.* **2018**, *33*, 095016. [CrossRef]
6. Baltz, V.; Manchon, A.; Tsoi, M.; Moriyama, T.; Ono, T.; Tserkovnyak, Y. Antiferromagnetic spintronics. *Rev. Mod. Phys.* **2018**, *90*, 015005. [CrossRef]
7. Cheng, R.; Xiao, D.; Brataas, A. Terahertz Antiferromagnetic Spin Hall Nano-Oscillator. *Phys. Rev. Lett.* **2016**, *116*, 207603. [CrossRef]
8. Khymyn, R.; Lisenkov, I.; Tiberkevich, V.; Ivanov, B.A.; Slavin, A. Antiferromagnetic THz-frequency Josephson-like Oscillator Driven by Spin Current. *Sci. Rep.* **2017**, *7*, 43705. [CrossRef]
9. Sulymenko, O.; Prokopenko, O.; Tiberkevich, V.; Slavin, A.; Ivanov, B.; Khymyn, R. Terahertz-Frequency Spin Hall Auto-oscillator Based on a Canted Antiferromagnet. *Phys. Rev. Appl.* **2017**, *8*, 064007. [CrossRef]
10. Sulymenko, O.R.; Prokopenko, O.V.; Tyberkevych, V.S.; Slavin, A.N. Terahertz-Frequency Signal Source Based on an Antiferromagnetic Tunnel Junction. *IEEE Magn. Lett.* **2018**, *9*, 1–5. [CrossRef]
11. Khymyn, R.; Tiberkevich, V.; Slavin, A. Antiferromagnetic spin current rectifier. *AIP Adv.* **2017**, *7*, 055931. [CrossRef]
12. Gomonay, O.; Jungwirth, T.; Sinova, J. Narrow-band tunable terahertz detector in antiferromagnets via staggered-field and antidamping torques. *Phys. Rev. B* **2018**, *98*, 104430. [CrossRef]
13. Safin, A.; Puliafito, V.; Carpentieri, M.; Finocchio, G.; Nikitov, S.; Stremoukhov, P.; Kirilyuk, A.; Tyberkevych, V.; Slavin, A. Electrically tunable detector of THz-frequency signals based on an antiferromagnet. *Appl. Phys. Lett.* **2020**, *117*, 222411. [CrossRef]
14. Ross, P.; Schreier, M.; Lotze, J.; Huebl, H.; Gross, R.; Goennenwein, S.T.B. Antiferromagentic resonance detected by direct current voltages in MnF$_2$/Pt bilayers. *J. Appl. Phys.* **2015**, *118*, 233907. [CrossRef]
15. Li, J.; Wilson, C.B.; Cheng, R.; Lohmann, M.; Kavand, M.; Yuan, W.; Aldosary, M.; Agladze, N.; Wei, P.; Sherwin, M.S.; et al. Spin current from sub-terahertz-generated antiferromagnetic magnons. *Nature* **2020**, *578*, 70–74. [CrossRef]
16. Vaidya, P.; Morley, S.A.; van Tol, J.; Liu, Y.; Cheng, R.; Brataas, A.; Lederman, D.; del Barco, E. Subterahertz spin pumping from an insulating antiferromagnet. *Science* **2020**, *368*, 160–165. [CrossRef]
17. Boventer, I.; Simensen, H.; Anane, A.; Kläui, M.; Brataas, A.; Lebrun, R. Room-Temperature Antiferromagnetic Resonance and Inverse Spin-Hall Voltage in Canted Antiferromagnets. *Phys. Rev. Lett.* **2021**, *126*, 187201. [CrossRef]

18. Lebrun, R.; Ross, A.; Gomonay, O.; Baltz, V.; Ebels, U.; Barra, A.L.; Qaiumzadeh, A.; Brataas, A.; Sinova, J.; Kläui, M. Long-distance spin-transport across the Morin phase transition up to room temperature in ultra-low damping single crystals of the antiferromagnet α-Fe$_2$O$_3$. *Nat. Commun.* **2020**, *11*, 6332. [CrossRef]
19. Cheng, R.; Xiao, J.; Niu, Q.; Brataas, A. Spin Pumping and Spin-Transfer Torques in Antiferromagnets. *Phys. Rev. Lett.* **2014**, *113*, 057601. [CrossRef]
20. Popov, P.; Safin, A.; Kirilyuk, A.; Nikitov, S.; Lisenkov, I.; Tyberkevich, V.; Slavin, A. Voltage-Controlled Anisotropy and Current-Induced Magnetization Dynamics in Antiferromagnetic-Piezoelectric Layered Heterostructures. *Phys. Rev. Appl.* **2020**, *13*, 044080. [CrossRef]
21. Consolo, G.; Valenti, G.; Safin, A.R.; Nikitov, S.A.; Tyberkevich, V.; Slavin, A. Theory of the electric field controlled antiferromagnetic spin Hall oscillator and detector. *Phys. Rev. B* **2021**, *103*, 134431. [CrossRef]
22. Meshcheryakov, A.A.; Safin, A.R.; Kalyabin, D.V.; Nikitov, S.A.; Mednikov, A.M.; Frolov, D.A.; Kirilyuk, A.I. Temperature tunable oscillator of THz-frequency signals based on the orthoferrite/heavy metal heterostructure. *J. Phys. Appl. Phys.* **2021**, *54*, 195001. [CrossRef]
23. Baryakhtar, I.V.; Ivanov, B.A. About nonlinear waves of magnetization of antiferromagnet. *Sov. J. Low Temp. Phys.* **1979**, *5*, 759–770.
24. Zvezdin, A.K. Dynamics of domain walls in weak ferromagnets. *Pis'ma Zh. Exp. Teor. Fiz.* **1979**, *29*, 605–610.
25. Andreev, A.F.; Marchenko, V.I. Symmetry and the macroscopic dynamics of magnetic materials. *Sov. Phys. Uspekhi* **1980**, *23*, 21–34. [CrossRef]
26. Ohlmann, R.C.; Tinkham, M. Antiferromagnetic Resonance in FeF$_2$ at Far-Infrared Frequencies. *Phys. Rev.* **1961**, *123*, 425–434. [CrossRef]
27. Hutchings, M.T.; Rainford, B.D.; Guggenheim, H.J. Spin waves in antiferromagnetic FeF$_2$. *J. Phys. Solid State Phys.* **1970**, *3*, 307–322. [CrossRef]
28. Khymyn, R.; Lisenkov, I.; Voorheis, J.; Sulymenko, O.; Prokopenko, O.; Tiberkevich, V.; Akerman, J.; Slavin, A. Ultra-fast artificial neuron: Generation of picosecond-duration spikes in a current-driven antiferromagnetic auto-oscillator. *Sci. Rep.* **2018**, *8*, 15727. [CrossRef]
29. Machado, F.L.A.; Ribeiro, P.R.T.; Holanda, J.; Rodríguez-Suárez, R.L.; Azevedo, A.; Rezende, S.M. Spin-flop transition in the easy-plane antiferromagnet nickel oxide. *Phys. Rev. B* **2017**, *95*, 104418. [CrossRef]
30. Baierl, S.; Mentink, J.; Hohenleutner, M.; Braun, L.; Do, T.M.; Lange, C.; Sell, A.; Fiebig, M.; Woltersdorf, G.; Kampfrath, T.; et al. Terahertz-Driven Nonlinear Spin Response of Antiferromagnetic Nickel Oxide. *Phys. Rev. Lett.* **2016**, *117*, 197201. [CrossRef]
31. Artemchuk, P.Y.; Sulymenko, O.R.; Louis, S.; Li, J.; Khymyn, R.S.; Bankowski, E.; Meitzler, T.; Tyberkevych, V.S.; Slavin, A.N.; Prokopenko, O.V. Terahertz frequency spectrum analysis with a nanoscale antiferromagnetic tunnel junction. *J. Appl. Phys.* **2020**, *127*, 063905. [CrossRef]
32. Sulymenko, O.; Prokopenko, O.; Lisenkov, I.; Åkerman, J.; Tyberkevych, V.; Slavin, A.N.; Khymyn, R. Ultra-fast logic devices using artificial "neurons" based on antiferromagnetic pulse generators. *J. Appl. Phys.* **2018**, *124*, 152115. [CrossRef]

MDPI

Article

Deposition of Crystalline GdIG Samples Using Metal Organic Decomposition Method

Hyeongyu Kim [1,†], Phuoc-Cao Van [2,†], Hyeonjung Jung [3,†], Jiseok Yang [1], Younghun Jo [4], Jung-Woo Yoo [3], Albert M. Park [1,*], Jong-Ryul Jeong [2,*] and Kab-Jin Kim [1,*]

[1] Department of Physics, Korean Advanced Institute of Science and Technology, Daejeon 34141, Korea; kimkkol@kaist.ac.kr (H.K.); jiseok@kaist.ac.kr (J.Y.)
[2] Department of Materials Science and Engineering, Chungnam National University, Daejeon 34148, Korea; caovanphuoc91@gmail.com
[3] Department of Materials Science and Engineering, Ulsan National Institute of Science and Technology, Ulsan 44919, Korea; llllll1151@unist.ac.kr (H.J.); jwyoo@unist.ac.kr (J.-W.Y.)
[4] Division of Scientific Instrumentation, Korean Basic Science Institute, Daejeon 34126, Korea; younghun@kbsi.re.kr
* Correspondence: bertpark@kaist.ac.kr (A.M.P.); jrjeong@cnu.ac.kr (J.-R.J.); kabjin@kaist.ac.kr (K.-J.K.)
† These authors contributed equally to this work.

Abstract: Fabrication of high quality ferrimagnetic insulators is an essential step for ultrafast magnonics, which utilizes antiferromagnetic exchange of the ferrimagnetic materials. In this work, we deposit high-quality GdIG thin films on a (111)-oriented GGG substrate using the Metal Organic Decomposition (MOD) method, a simple and high throughput method for depositing thin film materials. We postannealed samples at various temperatures and examined the effect on structural properties such as crystallinity and surface morphology. We found a transition in the growth mode that radically changes the morphology of the film as a function of annealing temperature and obtained an optimal annealing temperature for a uniform thin film with high crystallinity. Optimized GdIG has a high potential for spin wave applications with a low damping parameter in the order of 10^{-3}, which persists down to cryogenic temperatures.

Keywords: garnet ferrite; compensated ferrimagnet; metal organic decomposition

Citation: Kim, H.; Van, P.-C.; Jung, H.; Yang, J.; Jo, Y.; Yoo, J.-W.; Park, A.M.; Jeong, J.-R.; Kim, K.-J. Deposition of Crystalline GdIG Samples Using Metal Organic Decomposition Method. *Magnetochemistry* **2022**, *8*, 28. https://doi.org/10.3390/magnetochemistry8030028

Academic Editor: Atsufumi Hirohata

Received: 31 December 2021
Accepted: 22 February 2022
Published: 27 February 2022

Publisher's Note: MDPI stays neutral with regard to jurisdictional claims in published maps and institutional affiliations.

1. Introduction

High frequency dynamics from the antiferromagnetic exchange are essential for achieving fast computing spintronic devices [1,2]. On the other hand, to utilize the antiferromagnetic dynamics, a method to overcome its lack of responsiveness to the external field needs to be devised. Compensated ferrimagnet, composed of two antiferromagnetically coupled sublattices [3], enables the fast dynamics of antiferromagnetic materials while having accessibility via external fields similar to ferromagnetic ones.

The spin wave carries information as a collective precessional motion of localized magnetization. Unlike electric current mediated by the motion of charge carriers, the spin wave does not involve joule heating when propagating. Thus, it has high potential in the applications such as low-power computing devices [3]. However, a low damping constant is required to ensure a long transport distance for practical application. Rare earth iron garnet (REIG) is one of the best candidates in ferrimagnetic magnon transport. As garnet ferrite is an insulator, there is no electron spin sink contribution to damping, which results in low damping [4].

One of the obstacles in using REIG is a complicated fabrication process [5], which requires a dedicated vacuum chamber [6] or annealing temperature near the melting point [6–8]. The metal organic decomposition (MOD) method provides a simple and reproducible method to deposit REIG. Unlike other methods, the MOD method only needs a spin coating of the inorganic compound solution and the annealing temperature lower than the melting point.

In this work, we prepared gadolinium iron garnet (GdIG) thin film on gadolinium gallium garnet (GGG) substrate using the MOD method. X-ray diffraction (XRD), transmission electron microscopy (TEM), superconducting quantum interference device vibrational sample magnetometer (SQUID VSM), and Atomic force microscope (AFM) were applied to check crystallinity, magnetic properties, and morphology of samples. We found that high-quality garnet samples with reproducible magnetic properties can be deposited, which can be used to study physical phenomena such as longitudinal spin Seebeck effect [9] and spin Hall effect-driven various magnetoresistances [10,11].

2. Materials and Methods

The precursor solution is prepared by dissolving inorganic compounds $Gd(NO_3)_3 \cdot 6H_2O$ with 99.99% purity (from Alfa Aesar, Ward Hill, MA, USA) and $Fe(NO_3)_3 \cdot 6H_2O$, which has 99.9% purity (from Sigma Aldrich, St. Louis, MO, USA), into the solvent. A mixture of 99.8% purity dimethylformamide (DMF/Sigma Aldrich, St. Louis, MO, USA) and 95% purity polyvinylpyrrolidone (PVP) was applied as a solvent [12]. DMF and PVP were chosen because the mixture can dissolve inorganic compounds and, hence, can be utilized to grow garnet samples using MOD method [12–14]. As the mixing rate of DMF and PVP can dominantly affect the thickness and morphology of the sample, we used the reported values for YIG and BiYIG [13] because they have similar structures with GdIG. The stoichiometry of the Fe compound and Gd compound was set to 3:5, based on the formula unit of $Gd_3Fe_5O_{12}$. The total concentration of metallic compounds was fixed to 17%.

In order to avoid the contamination at surface of substrate, the GGG (111) substrate was sonicated first in acetone and then in ethanol for 30 min each. After that, the sample was treated under Ar plasma (20 sccm of Ar, 100 @ power with 70 kHz) for 30 min. A MOD solution was spin-coated onto GGG (111) substrate at 500 rpm for 5 s and 3000 rpm for 30 s. The coating and solution preparing procedure was optimized to deposit 10 nm of garnet sample [12]. For thicker samples, we repeated the growth procedure multiple times. After coating, the sample was dried for 30 min at 100 °C using a hot plate. The sample was then annealed for 1 h under an oxygen-rich atmosphere at various temperatures (750 °C, 800 °C, 900 °C, and 1000 °C) to determine the optimal annealing temperature for high crystallinity and good surface morphology. A fixed ramping rate of 4 °C/min was used to reach the target temperature.

Crystallinity was confirmed using a high-resolution XRD (RIGAKU corporation) with K_α emission line of Copper (λ = 1.54 Å). Surface morphology was examined by using AFM (XE-7 designed by Park Systems). Magnetic property, especially the magnetization compensation temperature (T_M), was measured by SQUID VSM in the Quantum Design magnetic property measurement system (MPMS).

3. Results
3.1. Structural Properties

To find the optimal post-annealing temperature for crystal formation, we first deposited a GdIG sample, targeting 10 nm thickness on the GGG substrate. Figure 1a,b show the measured XRD spectrum of deposited GdIG samples for various annealing temperatures. GGG (444) peak appears at 51.2°, which indicates that the lattice constant of GGG is 12.37 Å. Next to GGG (444) peak, a broad peak is observed at 50.3°, which corresponds to the GdIG (444) peak with a lattice constant of 12.55 Å. We note that the lattice constant of GdIG on GGG is larger than that of a single crystalline GdIG of 12.48 Å due to the pseudo-morphic growth on the GGG substrate [15]. We confirmed that there is no extra peak in the entire XRD pattern (Figure 1a) except for 111 orientation, suggesting the high epitaxial quality of GdIG. The thickness of GdIG was confirmed using a fringe pattern in the XRD spectrum. For the 750 °C annealed sample, the first and second fringe peaks were observed at 52.08° and 52.98°, from which the thickness of GdIG was estimated to be 11 ± 2 nm. This confirms that the thickness of GdIG is almost the same as what we targeted.

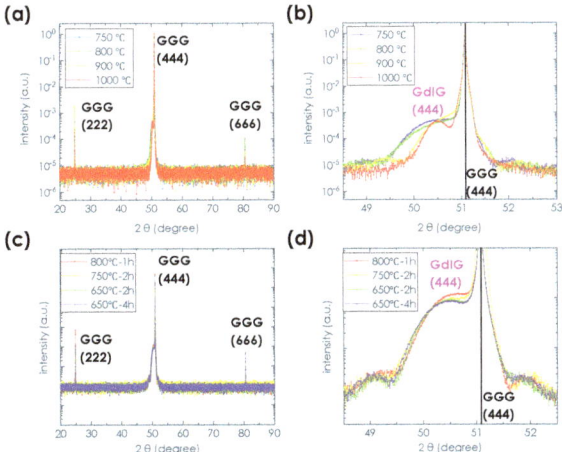

Figure 1. XRD spectrum for GdIG samples annealed at various annealing temperatures (**a**) from 20° to 90° and (**b**) from 48.5° to 53°. XRD spectrum for GdIG samples annealed at various annealing times (**c**) from 20° to 90° and (**d**) from 48.5° to 53°.

As shown in Figure 1b, the increased annealing temperature drives GdIG peak shaper. As the broadening of XRD peak is caused by vacancy- or defect-driven lattice expansion, the XRD result suggests that the higher annealing temperature results in better crystallinity [5]. We also confirmed that annealing time does not exert any significant change in the XRD pattern (Figure 1c). This suggests that crystallinity is predominantly governed by the annealing temperature.

However, we find that the high annealing temperature can affect the surface topography. The surface topography map measured using AFM in the non-contact mode (Figure 2) shows a significant change in surface morphology as annealing temperature increases. The roughness of film surface is about 0.1 nm order for the 750 °C and 800 °C annealed samples (see Table 1). However, the surface morphology of GdIG changes abruptly above 900 °C, and the island starts to appear. For 1000 °C annealed samples, island growth becomes the dominant film formation mechanism, resulting in the entire area being covered with islands of ~200 nm diameter and ~20 nm height.

Table 1. Roughness of films (Ra) for several annealing temperatures.

Annealing Temperature (°C)	Ra (nm)
750	0.14
800	0.08
900	0.71
1000	8.71

As the purpose of this work is to deposit uniform film for spin wave application, we selected 750 °C postannealed sample and performed additional measurement using TEM to ensure the crystallinity. TEM measurement was conducted using the STEM (scanning transmission electron microscope) system with Cs-corrector (JEM-2100F) designed by JEOL. A high-resolution TEM image in Figure 3a shows that GdIG and GGG have a smooth interface with a continuous crystal structure above and below the interface due to a similar lattice constant between the two materials. Additionally, a clear separation of Fe in the

GdIG film and Ga in the GGG substrate from energy dispersive X-ray spectroscopy (EDX) (Figure 3b) indicates minimal mixing between GdIG and substrate.

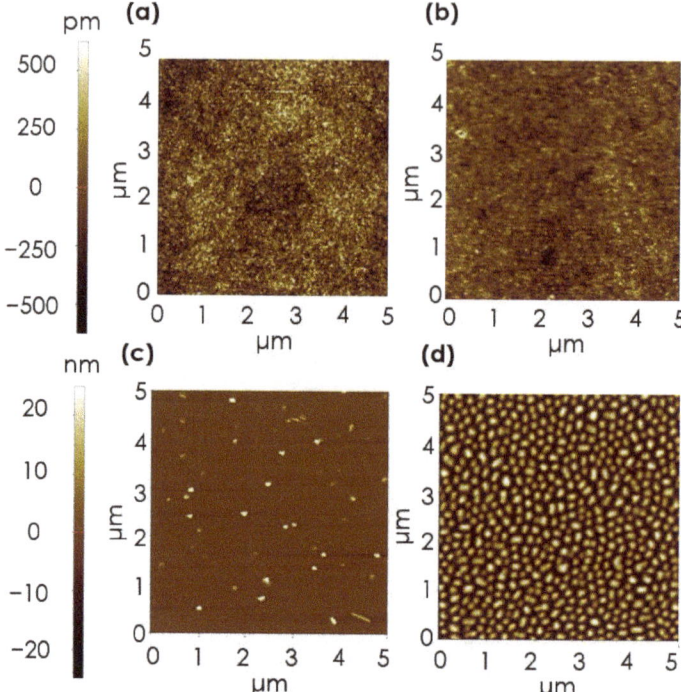

Figure 2. AFM image of (**a**) 750 °C, (**b**) 800 °C, (**c**) 900 °C, and (**d**) 1000 °C annealed samples.

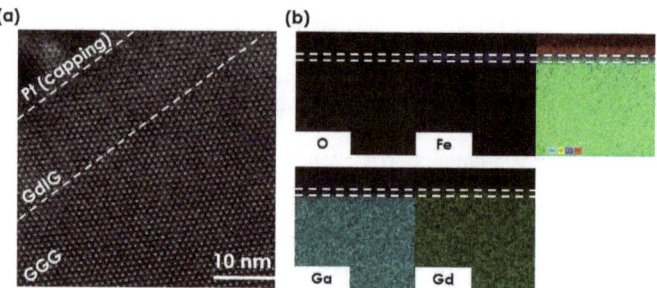

Figure 3. (**a**) TEM and (**b**) EDX images of 750 °C annealed GGG|GdIG sample. As GGG and GdIG have similar structure and lattice parameters, a smooth interface appears, indicating an epitaxial growth of GdIG. (**b**) The presence of Gd and Fe and the absence of Ga at the marked region represent the formation of the GdIG layer.

3.2. Magnetic Property

The magnetic property was checked as an additional measure to ensure crystallinity and stoichiometry. Temperature dependent magnetometry of 10 nm GdIG sample annealed at 750 °C was performed using SQUID VSM in MPMS 3. An M-T curve was obtained by measuring M-H curves under in-plane magnetic field at various temperatures and extracting the saturation magnetization (M_S). We eliminated paramagnetic contribution from GGG substrate by subtracting the linear background in the M-H curve at each temperature.

Figure 4 shows the extracted M-T curve. We focus on the value of M_S at low temperatures and the magnetic compensation point. M_S at 20 K was about 700 emu/cc, similar to previous reports as shown in Table 2. Magnetic compensation point appeared near 270 K also agrees with references on bulk and thin film GdIG [9,16,17].

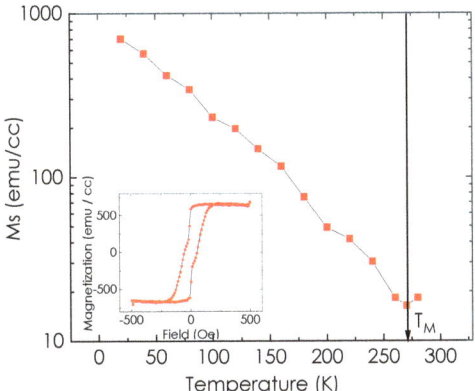

Figure 4. M-T curve of GdIG 10 nm thin film post-annealed at 750 °C. Inset shows the M-H curve at 20 K.

Table 2. Compensation temperature T_M and saturation magnetization (M_S) at 20 K of GdIG.

Source	T_M	M_S (emu/cc)
From Figure 4	270 K	700
From [16]	280 K	600
From [17]	300 K	670
From [9]	280 K	1300

3.3. FMR Measurement

Finally, the damping parameter of the GdIG sample was measured using the temperature dependent FMR. We used 60-nanometer-thick GdIG sample to enhance RF absorption power. Temperature and field control was performed using a physical property measurement system (PPMS). RF field for FMR was applied using a custom-built FMR setup that is compatible with PPMS.

Figure 5a shows the FMR spectra of the GdIG film for various temperatures. Here, we fixed the excitation frequency (15 GHz) and swept the magnetic field from 0 to 5000 Oe along the in-plane direction. Figure 5b shows resonance frequency as a function of external magnetic field for T = 50 K. The exact overlap with Kittel formula (red solid line) was observed. We extracted the damping parameter from the spectrum by measuring peak to peak distance from the FMR lineshape in Figure 5a. We first use the model for conventional ferromagnetic material to obtain damping parameter α_{FM}:

$$\Delta H = \Delta H_0 + \frac{4\pi\alpha_{FM}}{\sqrt{3}\gamma_{eff}} f$$

where $\gamma_{eff} = g_{eff}\mu_B/\hbar = \Delta M/\Delta S$ is the effective gyromagnetic ratio of the net moment, and ΔM and ΔS are the differences in the net moment and the spin density of two sublattices, respectively. Figure 5c shows ΔH as a function of frequency for several temperaures. The slope increases with increasing temperaure. Considering that the temperature variation of γ_{eff} is small, as seen in Table 3, the result suggests that α_{FM} increased with increasing temperature, which we ascribed to the divergence of α_{FM} at angular momentum compensation point (T_A) [18–20].

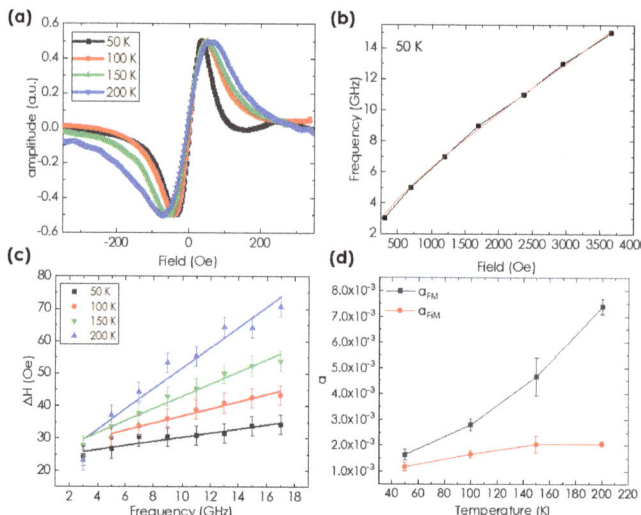

Figure 5. (**a**) FMR spectra of GdIG 60 nm samples at 15 GHz RF frequency. The horizonal axis (field axis) is shifted as much as the resonance field to compare the linewidth for different temperatures. (**b**) Resonance frequency as a function of the external field at 50 K and the Kittel formula fitting (red line). (**c**) Frequency dependence of peak-to-peak FMR linewidth. The solid line represents linear fitting. (**d**) Temperature dependence of damping parameters α_{FM} and α_{FiM}.

Table 3. Effective gyromagnetic ratio (γ_{eff}) and damping parameter (α_{FM}) measured at various temperatures below T_M.

Temperature (K)	γ_{eff} ($\times 10^7$ $T^{-1}s^{-1}$)	α_{FM} ($\times 10^{-3}$)
50	1.78 ± 0.08	1.60 ± 0.22
100	1.85 ± 0.07	2.80 ± 0.22
150	1.81 ± 0.02	4.67 ± 0.74
200	1.75 ± 0.08	7.41 ± 0.3

In the case of ferrimagnet, however, the dissipation rate from the two lattices and the dependence of resonance frequency on net spin density need to be considered to deduce a well-defined damping constant [18–21]. Since the GdIG sample also can be regarded as a compensated ferrimagnet with two effective sublattices, damping α_{FiM} is expressed as follows:

$$\alpha_{FiM} = \frac{|s_{net}|}{|s_{total}|} \alpha_{FM}$$

where $s_{net} = s_1 - s_2$ and $s_{total} = s_1 + s_2$ are the difference and the sum of the spin density for each sublattice ($s_{1,2}$), respectively.

To estimate α_{FiM}, we should extract the spin density of each sublattice in GdIG. This was achieved by using the M-T curve shown in Figure 4 and by assuming that the spin density of the Fe sublattice follows the same trend with that in YIG [16]. As shown in Figure 5d, α_{FiM} had no singular behavior close to angular momentum compensation in contrast to α_{FM}, and this dependence corresponds to previous reports on α_{FiM} of compensated ferrimagnets, including metallic GdFeCo [20].

We note that the GdIG sample deposited by the MOD method has comparable damping constant to some of the literature on spin wave applications [22,23]. In the case of the 10 nm YIG sample grown by the Pulsed Laser Deposition (PLD) method, an order of 10^{-3} damping parameter was obtained, which is similar to our result [24]. Our data also suggest that the MOD-deposited GdIG can be an alternative platform that could be used for low-temperature magnonic applications.

4. Conclusions

A high-quality GdIG thin film on GGG substrate was prepared using the MOD method. We found that a transition of the growing model of GdIG occurs at around 900 °C above which the film grows with an island pattern. The optimal postannealing temperature was decided based on the growth mode of the film and the crystallinity confirmed by the TEM. Temperature dependence of magnetization was measured using SQUID VSM and confirmed that T_M was measured at 270 K. FMR measurement showed that MOD deposited GdIG exhibits comparable damping constant to various deposition methods. Moreover, our experiment found that the damping constant of GdIG decreased at lower temperatures. We expect that the MOD method provides a simple and high-throughput procedure to deposit GdIG, which has potential advantages for magnonic applications.

Author Contributions: K.-J.K. and J.-R.J. planned and supervised the study. P.-C.V. deposited the GdIG sample; H.J. and J.-W.Y. performed the FMR measurement. J.Y. measured AFM; Y.J. measured MPMS; H.K. and A.M.P. analyzed the data from AFM, XRD, FMR, TEM, EDX, and MPMS and wrote the manuscript. All authors were involved in the discussion of the results and commented on the manuscript. All authors have read and agreed to the published version of the manuscript.

Funding: This research was supported by the National Research Foundation of Korea (NRF) funded by the Korean Government (MSIP) (grant numbers: 2016R1A5A1008184, and 2020R1A2C100613612) and by KAIST; funded Global Singularity Research Program for 2021; and by Samsung Research Funding & Incubation Center of Samsung Electronics under Project Number SRFC-MA2002-02. Y.J. acknowledges the support by KBSI grant (No. D210200).

Institutional Review Board Statement: Not applicable.

Informed Consent Statement: Not applicable.

Data Availability Statement: Derived data supporting the findings of this study are available from the corresponding author (A.M.P., J.-R.J. and K.-J.K.) on request.

Conflicts of Interest: The authors declare no conflict of interest.

Abbreviations

The following abbreviations are used in this manuscript:

MOD	Metal Organic Decomposition;
FMR	Ferromagnetic resonance;
REIG	Rare Earth Iron garnet;
YIG	Yttrium Iron garnet;
GdIG	Gadolinium Iron garnet;
GGG	Gadolinium Gallium Garnet;
AFM	Atomic Force Microscopy;
XRD	X-ray Diffraction;
TEM	Transmission of Electron Microscope;
EDX	Energy Dispersive X-ray spectroscopy;
PVP	PolyVinylPyrrolidone;
DMF	DiMethylFormamide;
SQUID	Superconducting Quantum Interference Vibrating System;
VSM	Vibrational Sample Magnetometer;
PPMS	Physical Property Measurement System;
MPMS	Magnetic Property Measurement System;
PLD	Pulsed Laser Deposition.

References

1. Keffer, F.; Kittel, C. Theory of antiferromagnetic resonance. *Phys. Rev.* **1952**, *85*, 329. [CrossRef]
2. Keffer, F.; Kaplan, H.; Yafet, Y. Spin Waves in Ferromagnetic and Antiferromagnetic Materials. *Am. J. Phys.* **1953**, *21*, 253. [CrossRef]
3. Kim, C.; Lee, S.; Kim, H.G.; Park, J.H.; Moon, K.W.; Park, J.Y.; Yuk, J.M.; Lee, K.J.; Park, B.G.; Kim, S.K.; et al. Distinct handedness of spin wave across the compensation temperatures of ferrimagnets. *Nat. Mater.* **2020**, *19*, 980–985. [CrossRef] [PubMed]
4. Zhang, S.; Li, Z. Roles of nonequilibrium conduction electrons on the magnetization dynamics of ferromagnets. *Phys. Rev. Lett.* **2004**, *93*, 127204. [CrossRef]
5. Cao Van, P.; Surabhi, S.; Dongquoc, V.; Kuchi, R.; Yoon, S.G.; Jeong, J.R. Effect of annealing temperature on surface morphology and ultralow ferromagnetic resonance linewidth of yttrium iron garnet thin film grown by rf sputtering. *Appl. Surf. Sci.* **2018**, *435*, 377–383. [CrossRef]
6. Gomi, M.; Tanida, T.; Abe, M. Rf sputtering of highly Bi-substituted garnet films on glass substrates for magneto-optic memory. *J. Appl. Phys.* **1985**, *57*, 3888. [CrossRef]
7. Lee, H.; Yoon, Y.; Kim, S.; Yoo, H.K.; Melikyan, H.; Danielyan, E.; Babajanyan, A.; Ishibashi, T.; Friedman, B.; Lee, K. Preparation of bismuth substituted yttrium iron garnet powder and thin film by the metal-organic decomposition method. *J. Cryst. Growth* **2011**, *329*, 27–32. [CrossRef]
8. Uchida, T.; Watanabe, S.; Uchiyama, T.; Tachiki, T. Fabrication of STO buffer films on MgO substrates by the MOD method. *J. Phys. Conf. Ser.* **2008**, *97*, 012057. [CrossRef]
9. Geprägs, S.; Kehlberger, A.; Coletta, F.D.; Qiu, Z.; Guo, E.J.; Schulz, T.; Mix, C.; Meyer, S.; Kamra, A.; Althammer, M.; et al. Origin of the spin Seebeck effect in compensated ferrimagnets. *Nat. Commun.* **2016**, *7*, 10452. [CrossRef]
10. Liu, G.; Wang, X.; Luan, Z.Z.; Zhou, L.F.; Xia, S.Y.; Yang, B.; Tian, Y.Z.; Guo, G.; Du, J.; Wu, D. Magnonic Unidirectional Spin Hall Magnetoresistance in a Heavy-Metal–Ferromagnetic-Insulator Bilayer. *Phys. Rev. Lett.* **2021**, *127*, 207206. [CrossRef]
11. Dong, B.W.; Cramer, J.; Ganzhorn, K.; Yuan, H.Y.; Guo, E.J.; Goennenwein, S.T.B.; Klaui, M. Spin Hall magnetoresistance in the non-collinear ferrimagnet GdIG close to the compensation temperature. *J. Phys. Condens. Matter* **2018**, *30*, 035802. [CrossRef]
12. Thi, T.N.; Van, P.C.; Viet, D.D.; Quoc, V.D.; Ahn, H.; Cao, V.A.; Kang, M.G.; Nah, J.; Park, B.G.; Jeong, J.R. Morphology-dependent spin Seebeck effect in yttrium iron garnet thin films prepared by metal-organic decomposition. *Ceram. Int.* **2021**, *47*, 16770–16775. [CrossRef]
13. Zhang, D.; Jin, L.; Zhang, H.; Yang, Q.; Rao, Y.; Wen, Q.; Zhou, T.; Liu, C.; Zhong, Z.; Xiao, J.Q. Chemical epitaxial growth of nm-thick yttrium iron garnet films with low Gilbert damping. *J. Alloys Compd.* **2017**, *695*, 2301–2305. [CrossRef]
14. Dongquoc, V.; Kuchi, R.; Van, P.C.; Surabhi, S.; Lee, S.W.; Kim, D.; Jeong, J.R. Enhancing magneto-optical and structural properties of Bi-YIG thin film on glass substrate using poly[vinylpyrrolidone](PVP) assisted MOD method. *Ceram. Int.* **2019**, *45*, 20758–20761. [CrossRef]
15. Yang, B.; Xia, S.Y.; Zhao, H.; Liu, G.; Du, J.; Shen, K.; Qiu, Z.; Wu, D. Revealing thermally driven distortion of magnon dispersion by spin Seebeck effect in Gd$_3$Fe$_5$O$_{12}$. *Phys. Rev. B* **2021**, *103*, 054411. [CrossRef]
16. Liensberger, L.; Kamra, A.; Maier-Flaig, H.; Geprägs, S.; Erb, A.; Goennenwein, S.T.B.; Gross, R.; Belzig, W.; Huebl, H.; Weiler, M. Exchange-enhanced ultrastrong magnon-magnon coupling in a compensated ferrimagnet. *Phys. Rev. Lett.* **2019**, *123*, 117204. [CrossRef]
17. Becker, S.; Ren, Z.; Fuhrmann, F.; Ross, A.; Lord, S.; Ding, S.; Wu, R.; Yang, J.; Miao, J.; Kläui, M.; et al. Magnetic Coupling in Y$_3$Fe$_5$O$_{12}$/Gd$_3$Fe$_5$O$_{12}$ Heterostructures. *Phys. Rev. Appl.* **2021**, *16*, 014047. [CrossRef]
18. Isaac, N.; Ruizi, L.; Zheyu, R.; Se kwon, K.; Qiming, S. Survey of temperature dependence of the damping parameter in the ferrimagnet Gd$_3$Fe$_5$O$_{12}$. *IEEE Trans. Magn.* **2022**, *58*, 1–13. [CrossRef]
19. Kim, D.H.; Okuno, T.; Kim, S.K.; Oh, S.H.; Nishimura, T.; Hirata, Y.; Futakawa, Y.; Yoshikawa, H.; Tsukamoto, A.; Tserkovnyak, Y.; et al. Low Magnetic Damping of Ferrimagnetic GdFeCo Alloys. *Phys. Rev. Lett.* **2019**, *122*, 127203. [CrossRef]
20. Okuno, T.; Kim, S.K.; Moriyama, T.; Kim, D.H.; Mizuno, H.; Ikebuchi, T.; Hirata, Y.; Yoshikawa, H.; Tsukamoto, A.; Kim, K.J.; et al. Temperature dependence of magnetic resonance in ferrimagnetic GdFeCo alloys. *Appl. Phys. Express* **2019**, *12*, 093001. [CrossRef]
21. Kim, K.J.; Kim, S.K.; Hirata, Y.; Oh, S.H.; Tono, T.; Kim, D.H.; Okuno, T.; Ham, W.S.; Kim, S.; Go, G.; et al. Fast domain wall motion in the vicinity of the angular momentum compensation temperature of ferrimagnets. *Nat. Mater.* **2017**, *16*, 1187–1192. [CrossRef]
22. Jungfleisch, M.B.; Chumak, A.V.; Kehlberger, A.; Lauer, V.; Kim, D.H.; Onbasli, M.C.; Ross, C.A.; Kläui, M.; Hillebrands, B. Thickness and power dependence of the spin-pumping effect in Y$_3$Fe$_5$O$_{12}$/Pt heterostructures measured by the inverse spin Hall effect. *Phys. Rev. B* **2015**, *91*, 134407. [CrossRef]
23. Jungfleisch, M.B.; Zhang, W.; Jiang, W.; Chang, H.; Sklenar, J.; Wu, S.M.; Pearson, J.E.; Bhattacharya, A.; Ketterson, J.B.; Wu, M.; et al. Spin waves in micro-structured yttrium iron garnet nanometer-thick films. *J. Appl. Phys.* **2015**, *117*, 17D128. [CrossRef]
24. Jermain, C.L.; Aradhya, S.V.; Reynolds, N.D.; Buhrman, R.A.; Brangham, J.T.; Page, M.R.; Hammel, P.C.; Yang, F.Y.; Ralph, D.C. Increased low-temperature damping in yttrium iron garnet thin films. *Phys. Rev. B* **2017**, *95*, 174411. [CrossRef]

MDPI
St. Alban-Anlage 66
4052 Basel
Switzerland
Tel. +41 61 683 77 34
Fax +41 61 302 89 18
www.mdpi.com

Magnetochemistry Editorial Office
E-mail: magnetochemistry@mdpi.com
www.mdpi.com/journal/magnetochemistry

www.ingramcontent.com/pod-product-compliance
Lightning Source LLC
LaVergne TN
LVHW070541100526
838202LV00012B/338